THE
WILLING

JOSEPH HENRY WILKINSON

1st WORLD
PUBLISHING

THE WILLING

JOSEPH HENRY WILKINSON

Published by 1stWorld Publishing
1100 North 4th St. Fairfield, Iowa 52556
tel: 641-209-5000 • fax: 641-209-3001
web: www.1stworldpublishing.com

First Edition

LCCN: 2007935537
SoftCover ISBN: 978-1-4218-9815-5
HardCover ISBN: 978-1-4218-9816-2
eBook ISBN: 978-1-4218-9817-9

ABOUT THIS BOOK

Some say it's delightful
Most say it's insightful
All say, though not frightful

It's a self-helping mouthful
It's a road map with guideposts
That uncovers the life force
Which leads to the state that
We've come to call Happy

CONTENTS

Part One

Part Two

Part One

Chapter One

THE WILLING

It was another cold Nebraska winter night, the wind howled across the prairie and the snow piled up on the tumble weeds lodged in the fences. Inside an old humble farm house, Dr. Jayson, an archeologist, is an internally conflicted man, conflicted between what he thinks he understands and what he does not understand. He is a loving father and husband. He has just settled himself in a big overstuffed leather chair, which had been handed down to him from his Grandfather. Tomorrow he would be leaving for a two month archeological dig in Peru near the border of Equador. 'Who knows what mysteries he would uncover' and what would he profit and what would he lose. After gazing at the fire for a long while he began to scan his bookcase. Then feeling that something was amiss his eye caught sight of a book he had not seen there before. He got up and walked over to the book and pulling it free from the book case and after shaking off the dust he sat back down. He noticed the book looked old [but?] yet new! This was very puzzling to him and after propping up his feet and pulling a blanket over his legs to keep out the winters chill while the fire place blazed away and one thousand ballets played as the shadows of the flickering flames danced upon the walls accompanied by the symphony of the wind as it moved around the house and through the trees outside. He opened the book and after stareing at what looked like blank pages, 'suddenly' the words began to appear…...

Dear reader,

I want to tell you a story that surely happened in the not too far distant future. My name is called the willing master.

This is my message from the land of the Willing. 48,000 B.C

You will understand that the world will change when enough people read this book and if you read this book it could be YOU who tips the scale and changes the world!

Are you up for it? If you say yes, then continue to read but please be advised that you will need certain tools if you are going to change the world. The first tool is courage. Do you know what courage is?

courage is that non-physical inner strength we should use when necessary. We especially require courage to SPEAK the TRUTH even if unpleasant things might follow. I say that last part just to be fair. If we are going to be courageous we need to know what TRUTH is so we can stand for it. Let's look at the word truth.

Humankind has searched for the TRUTH since before written history, and has tried every possible way to find it.

Was all humankind looking in the wrong place? Only now, from the future, can I tell you the TRUTH.

But before I do I am going to ask you to do this exercise. Take three deep breath rounds of air very slowly on the count of your heartbeat. Breathe in always through your nose, fill the stomach and then the three sections of lungs, lower, middle and upper. Count out eight heartbeats. Now hold your breath for four heartbeats. Now breathe out slowly let the chest fall on the count of eight heartbeats and then hold for count of four heartbeats. Each set should take about 20-26 seconds.——I am waiting for you to do it. This is your first act of COURAGE.——

Thank you and welcome to the land of THE WILLING. I am truly honored to be in your presence. I will tell you a little later about the TRUTH but first I want to ask you something. What do you want to be when you grow up? Well you may not have thought much about that. Many people older than you would misunderstand that question in various ways. They might say they want to be rich or famous, be a doctor or lawyer, mayor, governor or a football player, or maybe they want to be a barber, a farmer, a nurse, a petroleum engineer or a teacher (that's my favorite).

But people much older, like your grandparents, will always say, "Do what will make you happy." (So there you have it, a perfect example of the TRUTH.) It matters not what you do. Happy is what you want "to be" when you grow up. [After you read this book, if you have the courage you will be HAPPY!]

So if some one older than you asks what you want to do when you grow up, you might say "I'm not sure, maybe I could teach", then smile. If they ask you what you want "to be" when you grow up, just smile and say, "Happy", then keep smiling.

Imagine a road connecting the City of COURAGE to the rapidly growing Town of HAPPY and between is a bridge called TRUTH. That pesky word pops up again and I did promise to tell you exactly, "What in the world is the TRUTH?" Could it be that all of mankind, all of these years, has been asking the wrong question? We hear of people going out into the world in search of the TRUTH. Instead of "what", let's change it to "who". Now we have the real question: "Who is the TRUTH?" The answer is obvious when you think about it. This is a good time to breath two good slow ones. I'll wait.---8-4-8-4, three times thru the nose and after the 2nd breath of air, look always for a feeling of calm; for when you are calm, your are in charge. You are in charge of your emotional field and the mind is quiet. When you are calm, you can think clearly and work without mistakes. You will be a better student and a better person. Older people call this calm feeling "Peace of Mind." When you are breathing normally your breath should be deep and slow and connected one the other. The 8-4's are for extra help in times where clarity is needed, or you are in need of gaining control of your emotions. We must learn to control the emotions always. We can never let the emotions control us.

COURAGE is your first tool and your breath is the second tool and HAPPY is where you want to go. Now we have only to find out the truth so that we can use our COURAGE to stand for it. Otherwise COURAGE is a hot air balloon that falls to the ground with a bump after the fuel runs out. COURAGE is nothing if it does not stand for the TRUTH.

Dear Reader, the TRUTH is: You are potentially Divine and along with your understanding of Wholeness and Unity will become the TRUTH. There really is only one TRUTH and you are part of it. The reader of this book is part of the TRUTH and anyone who may be listening is part of the TRUTH.

We are all like a very nice wooden table with a layer of dust on top. With a polishing cloth and a little effort the TRUTH will shine through for all to see. The TRUTH can best be realized if one is HAPPY.

Just remember you can be HAPPY if you are WILLING. Therefore you will change your world and when you change your world, the world itself will change.

The following pages concern a story that happened in the not too far distant future. Read carefully.

Chapter Two

THE DRAMA

Dr. Jason who is known to all as Dr. J, is an archeologist who planned to leave for a dig in the mountains of Peru. But just before he was scheduled to leave, a chain of events disrupted a whole afternoon. His eleven year old daughter, Patty, had just finished a peanut butter sandwich. "Dee-licious" she thought. She one fingered some jelly off the plate and into her mouth, she closed her brown eyes, savoring the last bit of grapey sweetness. Just then, disharmony broke out like a sack of groceries with a wet bottom, when Patty heard her mother's really upset voice. "Patty! Patty, if you can hear me, come into the kitchen and clean up your mess," yelled Sue. Sue did not like to yell. It made her feel out of control and even more upset than she already was. However, this time must surely count as an exception since Patty had left the lids off both the peanut butter and the grape jelly jars. The butter knife, left out on the counter, still had some peanut butter and dripping jelly on it. Sue started to lick the knife but decided to hold it for evidence.

"I'll be right there" shouted Patty as she bounded down the stairs and across the carpet leading to the kitchen. Just then her thirteen year old brother Phillip, sitting at the dining room table, his black hair shinning from a recent shampoo, said, "Hey Patty, come look at the photos of Sunday's picnic!" "Wow! Let me see, let me see!," Patty said excitedly, forgetting for the moment the call from the kitchen. Patty looked as quickly as possible through about half of the 40 or so printouts when Philip started teasing her. One particular picture showed Patty making a wrinkled face, trying to avoid a dog's attempt to lick her on the mouth. Not a complimentary picture to say the least. Then her older brother's teasing made her scream at him as she wrestled him for the picture.

As if right on cue, Phillip's elbow caught mom's favorite vase, knocking it to the floor. It broke into several pieces and the loud crash brought Sue into the dinning room.

Sue started yelling at Patty again for not coming immediately when called (yelled at). Then she saw the broken vase (said to be quite valuable). The vase had been handed down to her from her great-grandmother. When Patty said, "Phillip did it Mom", Sue turned to Philip to continue her tirade but she suddenly began to cry. Patty, seeing her mom sobbing like that, went over and hugged her and started crying too. Philip, embarrassed by what he had caused, went over and hugged his mom's other side. Sue, standing with her face buried in her hands, still attempted to yell through her tears.

Here is only one example, dear reader, of a family acting in disharmony. Had they known how to create harmony, they surely would have given it a chance to happen. From the aforementioned drama, you get the idea that Patty loves Sue and Phillip. Phillip loves Patty and Sue. But no evidence in the drama shows that Sue loves Phillip and Patty. It turns out, she does very much love them both. So what is wrong with this picture? [In my opinion, at this stage of social development, love is not enough. It's important to know how to interact.] Do you know how to behave? Think about it. Take three slow ones here. Take your time 20-25 seconds for each breath. I'll wait. As you can see I am serious about this breath thing. The more carbon dioxide you breathe out, the more oxygen can get to the brain to spark good things like laughter and creativity. Oxygen fuels the mind and body. The oxygen we breathe in must dissolve in hemoglobin, which is found in the bloodstream. Hemoglobin is like a big bus in a city where people get on and off. In this case, oxygen and carbon dioxide get on and off of the hemoglobin "bus". However most of our breathing is so shallow that not much oxygen enters the blood. Then there is the problem of carbon dioxide (CO_2) getting out of the bloodstream. CO_2 is offloaded in proportion to the air available, hence if only a little air comes in, it means only a little offloading of the CO_2, thus alas, very little life-giving, creativity-causing, happiness-helping, oxygen can dissolve in the bloodstream. The big bus cannot take oxygen passengers on until the CO_2 gets off. Excess CO_2 is of no value to the body. CO_2 needs to be offloaded at the first chance. Now you understand the importance of doing a good job of breathing. Everyone around you will see you as a better friend. And that is a good thing. Oh yes, back to the question "Do you know how to behave?" That is really what this book is about. Correct breathing is a very important act. A big part of life should be just having fun. And we will have lot of good fun as soon as we learn how to behave.

Chapter Three

SCENE II ACT 1 OF THE DRAMA...

Dr. J arrived home, and after parking the car, walked into the back yard and started a conversation with the next door neighbor. The neighbor complained that he was having a hard time keeping out of the way of his wife since he retired after 35 years at the Post Office... Dr. J suggested that he volunteer by getting involved in some community activities or maybe he should take up golf or coach a soccer team. Just then they heard a commotion coming from the house. Dr. J excused himself, while his older neighbor said he knew how it was with young families. Dr. J ran across the back yard faster and faster as the yelling, then crying, and yelling, sounded throughout the whole neighborhood. The neighborhood had plenty of trees, green grass and lots of flowers, some of which smell really wonderful, especially in the early morning. Oddly, Dr. J considered the fact that he had not smelled the flowers in his own yard for some time because he had been too busy, (Which was one of the excuses he used most often.) Dr. J leaped over the 2 foot tall flower bed in an effort to short cut his path to the house. The flowers blocked Dr. J's view of the ground on the other side of the flower bed where he intended to land. And land he did, right on the tail of Sphinx, the family cat. Sphinx, a large orange

tabby with a white streak down his nose and a big fluffy tail, was taking a nap in the shade of the flowers. Sphinx had long ago been neutered so he had plenty of extra time to lie near the catnip patch in the shade of the flowers.

Poor Sphinx didn't know what hit him. In the first minute he was relaxing (which is his favorite pursuit) in the after-glow of a freewheeling wallow in the catnip patch. The next minute he felt like someone just put a torch to his tail. His first reaction was to turn and claw the heck out of the one holding the torch. Dr. J knew immediately what happened when he heard Sphinx yowl as only a cat in trouble can do. Dr. J, in an effort to get off the cat's tail as quickly as possible, turned his ankle and fell to the ground with a thud with Sphinx's claws buried in his calf.

Poor Sphinx! Now that his tail was ok he realized that he had just made a big mistake. This was no ordinary calf, this calf was attached to someone who pets him and sometimes fills his food bowl. Sphinx disengaged his claws from Dr. J's calf and ran for cover.

Dr. J jumped up and glanced to see Sphinx diving into the bushes. (He didn't know Sphinx could run that fast) But Dr. J had lost none of his zeal for getting to the house as quickly as possible. He got back on his feet and with a noticeable limp, hobbled quickly across the yard toward the back door leading to the kitchen. He burst into the kitchen and headed in the direction of the yelling and crying. He caught a glimpse of the plastic trash container just before he booted it like an extra point in football high above the linoleum floor.

The trash container began to spin like a Ferris wheel spilling trash everywhere. Disbelieving, Dr. J looked as the trash container landed upside down on the kitchen table totally void of any evidence of its purpose. Dr. J, having booted the container full of trash with the same leg that had a turned ankle and a calf full of claw marks, now winced as pain shot up to his hip bone. Dr. J, now limping noticeably, continued in the direction of the disturbances. Just like when you are away from the phone, and it rings and rings, you run and you run to get it, then you can hear them hang up just as you breathlessly say hello. Well the noise stopped, just like that as Dr. J. entered the room and took one look. There was Sue, his wife and mother of their two children, Phillip and Patty, who were standing there hugging their mom.

He limped over to the dining room table and sat down and invited his family to join him. They all sat down and joined hands while Dr. J offered a short prayer asking Great Spirit to help the family gain some insight from what had happened in only 5 minutes time.

Everyone listened as Patty first told her side of the story. When she finished, everyone agreed that Patty had not acted in a responsible manner on two counts: first, she had not put away the peanut butter and grape jelly, and second, she should have told Philip that she could not look at the photos just yet because her mom wanted her in the kitchen.

Philip told his story and it was obvious that Phillip had heard his mom call Patty and it was irresponsible for him to tempt Patty with photos while she was headed for the kitchen to answer her mom's call. Phillip also acted wrongly by teasing Patty about the bad face in the photo with the dog, and thirdly, Phillip caused the tussle that resulted in the broken vase. Phillip said he would glue the vase back together as a part of the consequence of his behaviors.

Then it was Sue's turn to say what had happened. Sue told of some little things that happened that day: The milk she had asked Dr. J to pick up the previous afternoon did not happen. He had not brought milk home, so there was no milk for breakfast. Later the washing machine broke in the middle of the first of six loads of laundry, then the phone rang, and the neighbor who attended last Sunday's picnic said she needed her table cloth. Phillip had bumped the table and spilled the gravy bowl all over the tablecloth while being chased by Patty who was upset because Phillip had taken a picture of her trying to escape a dog's lick on the mouth. Sue had volunteered to wash it as soon as possible. Sue called the washing machine repair man and found out that he could not come for a couple of days. So Sue washed the table cloth by hand and returned it. All these little things built up great stress in her mind and body. Then as she was cleaning and collecting trash to carry out back, she walked through the kitchen and to the back door where she set the trash can down because she forgot the kitchen trash under the sink. When she went over to get the trash from under the sink she saw Patty's mess. (You know the rest.)

She felt terrible about the yelling and admitted that she felt overcome with frustration, not only because of what Patty had left undone but because the entire day had gone so badly.

Everyone agreed that yelling didn't help and should only be done in the case of extreme emergency. Sue agreed that leaving the trash can in the middle of floor had unforeseen consequences.

Next, Dr. J admitted that he had too many things on his mind and had forgotten to pick up the milk. He knew he should have been thinking more of his family and less about work. He then told the family of an archeological dig in which he was invited to participate. He told the family that he would leave in three days and that he was departing early for the scheduled dig. He planned to return in two months. Then Dr. J recounted his adventure. He told of his conversation with the neighbor, the leap over the flowerbed and all the rest.

Everyone laughed at his story even though it was of course not funny at the time. Even Dr. J laughed as he showed the claw marks from Sphinx and the slightly swollen ankle that he had turned. When he told of the booting of the trash can and the watching its majestic rise in the air and spinning like a Ferris wheel, spewing trash all around, and then landing upside down on the kitchen table, he had Patty and Phillip rolling on the floor with Sue holding her sides in laughter. Dr. J quietly wished the family could be more like this in the future. He realized that a family should be happy, not crying and yelling and wrestling and breaking things. But, as for now, this was just a wish for he had no answer as to how his family could be happy.

After the laughter subsided and order was restored, and with Patty and Phillip back in their chairs, the family rejoined hands, agreed to try their best to do better and avoid a similar situation in the future. Dr. J closed the circle by giving thanks for the clarity of the meeting. Dr. J. and Patty went to the kitchen to clean up while Philip went after some glue for the broken vase. Sue went for medical supplies for Dr. J.

Many hands make for short work. Dr. J, Sue and Patty had the kitchen back to shape in short order. Dr. J carried out the "picked up" plus the "under the sink" trash that Sue had forgotten earlier. After dumping the trash in the alley, Dr. J found Sphinx and apologized for the tail stomping. Sphinx seemed not to remember as he purred and eagerly accepted the rubbing and stroking. On his way back to the house, Dr. J stopped to smell the flowers. Back in the kitchen he helped with the evening meal. Patty went to her room to fetch the now empty peanut butter sandwich plate. Also Patty agreed to clean up her room a little better.

Soon Sue had the meal on the table. Dr. J had made some really great sauce while Sue had cooked spaghetti and made a fresh garden salad. Dr. J had asked earlier for Phillip to set the table which Phillip did. When Dr. J and Sue brought the food to the table they were both surprised to find that Phillip had also taken time to glue the vase back together. There it stood on the side table, as if nothing had happened. Dr. J high-fived Phillip, then Sue gave Phillip a big hug, told him how much she loved him, adding that she was sorry for yelling at him. Phillip said, "Thanks Mom, I will try to do better in the future". Phillip went to the foot of the stairs and called for Patty to come to the table for the evening meal. All was well, at least for awhile. Dr. J remembered, when he was young, that many books had the same ending "And they lived happily ever after." Now books never end that way. Is it because happiness is no longer an option. Is happiness no longer possible? Soon Dr. J would leave for Peru.

[Now is a good time to find out where in the world is Peru on the world map. Once you find Peru then locate the capital. Find out the altitude of the Capital City- - - - - I'll wait - - - -But first, see if you are naturally taking deeper stomach breaths. In case there is confusion here, It should be made clear that the 8-4's are to be used in time of stress or when you need to clam you mind such as for a test or speaking in front of a group. Your normal breathing should now be connected full and slow and calm. Much like circular motion. So check in once in a while and keep thinking about doing the practice. The Road to Happiness is paved with a well-practiced breath.]

Chapter Four

THE SHALLOW EARTHQUAKE

News of a shallow earthquake in a sparsely populated part of the jungle in Peru caught Dr. J's eye. The quake happened in the same part of the country as his upcoming dig. Dr. J sat in the International Airport in San Francisco reading a newspaper while waiting his departure for Lima, the capital city of Peru. (See! You knew that.)

The quite shallow earthquake seemed to have affected only a half square mile or so. Gee, thought Dr. J, a half of a square mile about 300 acres, about the size of an 18 hole golf course, driving range, with plenty of water hazards and some houses all around. Dr. J loved the game of golf, a game that he played as often as he could.

Golf is a lot more than a club and a ball. The process of play-ing the game gives one plenty of time to understand one's emotions. Dr. J had three or four buddies who always played together. They always had a good time and laughed a lot.

Dr. J realized his mind was wandering so he refocused on the news story. The story noted little interest in the scientific community to investigate, as there could be little damage done to those in the sparsely populated jungle. The story had a deep impact on Dr. J as he tried to think of a similar shallow earthquake anywhere or at any time. He was not scheduled to check in with the Archeology Department at the University for 15 days or so. He had left early because he felt he needed the time to learn about Peru and brush up on the language. Dr. J had taken a couple of Spanish classes in college. Spanish came in handy on digs in Mexico. South of the U. S A, beginning at the border of Mexico and thru Central and South America, the people all speak Spanish except Brazil where they speak Portuguese. Look up Costa Rica on the map. Do you know that Costa Rica does not pay any money for a standing army? Therefore, they are the only country in North and South America that does not have a standing army and it turns out Costa Rica is doing just fine. Costa Rica is doing its best to show the TRUTH about guns and bombs and jet planes with both.

[Have you noticed if you are breathing with the stomach-then upward, without thinking about it yet? Keep trying. Believe it or not, it will kick in. It is our 'birth-right' to know how to breathe-right. Once you learn how to be aware of breath like never before, you will have made a great start toward the "knowing" how to act part. You will need your COURAGE when you make the effort to absorb and apply what you learn in this book. [It is easy when you do it every-day but maybe not at first.]

On the plane, Dr. J put on the earphones and listened to some soft music. He studied his Spanish vocabulary book. All too soon, it seemed, he was alerted by the sound to fasten the seat belt for the landing at Peru International Airport.

He checked in at a hotel in downtown Lima and went for a walk to look for a place to eat dinner. After a leisurely walk, he stopped by a coffee shop just to hear local conversation and the local issues that they may discuss.

Sitting in a sunny spot out in front of the coffee shop, Dr. J saw two men come over and sit at a table near him Dr. J picked up much of the Spanish being spoken, although he still missed a word or two in the course of the conversation. Almost at once the two men started talking in geological terms while referring to the shallow earthquake that had recently occurred. He edged a little closer to hear and just then he caught the eye of one of the men. "Buenos Dias" said Dr. J; realizing that he had been too obvious in his attempt to get closer to the conversation. "Buenos Dias" the man said with a smile while the other man turning to see him also smiled. Dr. J. took the liberty to introduce himself as an Archaeologist. The two men seemed not to have a problem with Dr. J's Spanish. Soon they were talking excitedly about the earthquake and possible explorations of it. They told Dr. J neither the University of Peru nor anybody else wanted to go investigate. "Where could I find a guide if, for instance, I wanted to go see it for myself ?" D. J. blurted out these words before he even thought what he was thinking. Both men looked surprised at such interest and one said that he knew the best guide in all of Peru. "He could take you if he's not busy."

"It will take about four days one way to hike in. If you figure six days on the site, you could be back in two weeks." The two men, warm and friendly, seemed to smile a lot and Dr. J felt very comfortable as he sat there sipping on his third lime-lemon drink. He realized how good it felt to be around people who smiled a lot. One of the men stood to use the phone and came back shortly with news that the guide would be happy to take him to the site but it would take three days to get ready. He gave Dr. J the guide's name and phone number. Dr. J stuffed the paper in his pocket. Soon the men had to leave but not before they each invited him over for dinner. Dr. J had dinner the next two nights at both their homes. Dr. J ate very tasty wholesome food. The conversations were lively and informative but these folks had their family troubles too. It seemed no one knows how to be happy as a way of life.

Chapter Five

DR. J AND JOSÉ GO FOR A HIKE

On the first day after meeting the two men at the coffee shop, Dr. J had phoned the guide, José, to set up a supply list and payment schedule for the trip. His quoted price was reasonable and the supply list shortened because their route would pass thru villages with food on the way to the remote earthquake site.

So Dr. J. met Jose and after talking for while it was apparent that could get along together and do well. Carlos was average in height but his body was very strong in appearance. His legs looked like telephone poles almost with very well toned muscles.

The first day they drove as far as they could, then they started to hike. Dr. J thought that his back and legs were strong enough to make the hike, but the hike took five days instead of four, mainly because Dr. J needed to acclimate himself to the thin mountain air. Dr J understood the body's need for oxygen now more than ever. He saw the simple beauty of the high villages of Peru where the mountains to go as high as

20,000 feet. Climbing upward with a 60 lb. backpack and still breathing enough oxygen to satisfy his body became very difficult. His legs seemed to be on fire some of the time, but Dr. J was able to keep up mainly because there were plenty of rest stops. José was being paid to be guide to Dr. J, not to run off and leave him. By mid afternoon of the third day, they had reached the last village before entering the jungle and the guide suggested staying the evening in the village, eating and getting to bed early. Dr. J agreed. It would be good to spend some time with the villagers.

Chapter Six

THROWING THE BONES

José spoke to one of the village leaders about food and a place to sleep. They quickly settled into one of the empty huts and Dr. J, unloaded his backpack with great relief. José, the guide, was all smiles now that they had reached the jumping off spot to the largely uninhabited jungle. Dr. J and José walked around the village, occasionally stopping to talk with the women who sat on rugs outside their huts happily weaving rugs or garments. It seemed all of the women had smiles on their faces. When Dr. J questioned José about these unusual phenomena, José gave a quick and easy smile and explained that women chew on certain leaf that allows them to feel easy and happy all day long. José went on to say that the villagers seemed to live long happy lives up in these villages so high in the mountains. Dr. J looked over to the side and saw a very old woman sitting on a rug. She had large eyes that seemed full of wisdom. After a few steps, he glanced back to see the old woman was still looking at him. He felt compelled to walk over and talk to her. José was busy talking to some of the villagers. Dr. J sat down and introduced himself to the old woman who smiled and nodded her head. .She offered a cup of tea which he gladly accepted. As he sipped, the old woman watched him carefully. Dr. J asked, "How long have you lived in this village?" The old woman seemed not to hear the question as she said, "You have something that is bothering you deeply." He was surprised to hear such a statement coming from the old woman. After all, he didn't think he was all that transparent. However, being there in the village, high in mountains, miles from anywhere, he felt he might as well talk to the wise old woman "Yes" he said, "I have a problem with my family back in Nebraska."

The old woman lowered her head and in a quiet voice said, "I throw the bones. If you have a problem, perhaps the bones will help you solve it. But I cannot do it for free." Dr. J understood and they decided on a small amount which he paid up front. The old woman smiled and took out a small pouch and dumped the contents on the rug the contents looked like bird and small animal bones. She said, "So it seems you are concerned about the happiness of your family, is that right?" Again surprised, he replied, "Yes, that's right. I want to know how my family can be HAPPY."

The old woman's eyes focused on Dr. J and she seemed to understand his question. She picked up the bones and gave them a toss on the rug. After studying the bones for a while she said only a few words. Dr. J did not know what to expect since how to achieve happiness is a hard question. But when he heard her answer he was sure he'd wasted his money and time. The old woman intoned "You will find your answer in the side of a mountain." Dr. J finished his tea, thanked the old woman for her time and effort, than joined José, who by this time had a large group of villagers around him as he talked of places he'd been to in the largely unpopulated jungle.

Later Dr. J and José shared a nice meal blessed with the grace of the loving hands that provided it. He fell asleep easily that night knowing he needed rest for tomorrow's journey.

José rose before dawn and now sat with a cup of coffee and

some hot cereal in front of him. Our Archeologist opened his eyes, smelled evidence of cooked food. then moved quickly to get up but, when his sore legs talked back and he slowed significantly in his motion. He dressed and packed his sleeping bag then helped himself to two bowls of cereal and some coffee.

After saying their goodbyes to the villagers, José and Dr. J set out on their journey into the bush and largely unpopulated jungle. The passage to the area seemed remarkably easy because of the paths apparently already in place and leading generally towards the shallow earthquake. That evening they made camp under a giant tree with all manner of life in its limbs. Colorful birds with long tail plumes sang songs as if in competition but at nightfall the noise stopped as if signaled by an unseen timekeeper.

Dr. J made a fire, boiled water, added some freeze dried stew and they ate a delicious evening meal. As José drank his last cup of tea before bed, Dr. J asked how it happened that the paths they took seemed always to be in the general direction of the earthquake. José smiled and said "The jungle always provides. We believe the jungle is alive and, of course, so is the earth. Since they approve of this trip they help in the ways they can. Easy trails, good weather and freedom from harm, for example, are ways that the mother of all nature can display her power."

Dr. J thought on that and recalled that his Lakota Sioux father held the same understanding about mother earth and all the many gifts she bestows on those in harmony with her. He felt honored and encouraged that the Peruvians felt the same.

The next morning José seemed to move more slowly in breaking camp. When Dr. J questioned his casual pace, José smiled and said he had a dream that the path would be clear today and that they should reach the site by mid-afternoon.

Now Dr. J beamed. Hiking another full day with the backpack and his still sore muscles would have been quite a challenge. Dr. J poured the last of the coffee on the fire to quench it and dumped some dirt on it for good measure.

They plunged into the jungle. They had gone only several yards when they stepped onto a path leading straight toward the earthquake site. The wide path led Dr. J to assume maybe other people also used it, but José replied that hunting parties seeking game probably did use it. By mid-afternoon, true to José's dream, they reached the earthquake site. José made camp, while Dr J took a look around. The late afternoon was cloud-covered when suddenly the sun broke thru and lit up the side of the large mountain above Dr. J.

He looked up the mountain side and there, starting more than halfway up, he saw the evidence of a large land slide which had stripped the trees and foliage into a rather neat pile at the base. Dr. J looked at the rubble and then at the land surrounding. A short walk about indicated this was not an earthquake at all, because the area affected was so small. The only conclusion Dr. J could come up with seemed absurd. The shallow earthquake report now appeared to be an assumption not based on fact. Another hypothesis might be that an underground bomb-like explosion had occurred. Perhaps there was an ammunition dump, for a small guerrilla army in the area. The setting sun prompted him to make the hike back to camp. He moved quickly toward camp, all the while wondering what really happened here. So far nothing made sense.

Long before Dr. J arrived in camp he smelled enticing aromas from the campfire José had set up. Now that he had hiked four days, he noticed strength in his legs and he had a little less soreness. He strode even faster along the path to camp. His arrival came complete with a big appetite and he happily saw that José had finished cooking and was waiting for Dr. J's return. José had prepared Dr. J's favorite, a dish of yams and rice flavored with some enigmatic spice. He greeted José, fixed a plate and then handed it to José. José raised his eyebrows in surprise, gave a big broad smile as he accepted the plate. "Gracias, amigo" José said, giving Dr. J a cup of tea. "Gracias", said the equally smiling Dr. J as he served himself and began the meal.

José listened intently as Dr. J told him about his findings and his theory of the ammunition dump. José smiled and said the peaceful Peruvians really have no reason to take up arms. Dr. J gazed up the mountainside at the landslide and thought while his scientifically trained mind raced around for any explanation to this mystery.

Chapter Seven

THE MOUNTAINSIDE

After the sun went down and darkness crept in, Dr. J listened as José told him of the wonders of the jungle that he had seen first hand as a guide. José spoke of giant water falls, exotic and tempting foliage, and all manner of brilliantly-colored birds. He also spoke of things that back in Nebraska would be considered creepy and crawly. "What you see in jungle is life teeming in every form. There seems to be room for all life forms to thrive in harmony in the jungle."

Dr. J had been staring at the campfire for more than two hours while José talked. When José finished, he stood up, stretched his back, walked away from the light of the campfire. Looking up in the direction of the mountain he saw a bit of light on its side. Were his eyes playing tricks? Had he looked at the campfire so long? But even after blinking and refocusing he still saw the light. He called José over away from the camp fire and into the darkness. Then Dr. J asked José if he could see the light on the side of the mountain. José looked up and he said "I can see it! Dr. J. I can see it too, but I cannot tell you what is causing the light.".

Dr. J and José returned to the campfire, both excited about climbing the mountain to investigate at daybreak. Dr. J retrieved his binoculars and walked back out into the darkness for a closer look at the light's source. Focusing on the light, he could see clearly a cave opening possibly exposed by the mysterious landslide.

Dr. J stretched his back again as it had not recovered as fast as his legs. His mind wandered back to the village and the old woman who had read bones and told him happiness for his family would be found in the side of a mountain. Could it be? Dr. J headed for his sleeping bag, his mind still spinning with all the happenings since arriving at the site. He slipped into his bag and glanced over to see José already fast asleep. How did he do that so easily thought Dr. J, while hoping he could put his mind to rest so he too could get some sleep for the long uphill hike in the light of a new day.

Dr. J awoke at first light, quickly crawled out of the sleeping bag and walked over to the campfire where José had tea and oatmeal waiting for him. "Buenos Dias, José," Dr. J said smiling before his first sip of tea José gave him his broad smile and returned the greeting. "Buenos Dias, Dr. J I hope you are ready for long climb up the mountain." Dr. J winced a little at the thought of climbing up what looked like a 45-degree angle to the cave opening. José suggested they climb up the mountain through the trees rather try to walk in the potentially dangerous area where the land slide had occurred. Dr. J agreed with the plan and decided he would let José choose exactly where to start the climb up to the cave.

He ate a second helping of oatmeal and thought, not of the hard climb that awaited, but of the fact that the light was not always there for anyone to see because the last trail they took led almost to their chosen campsite, therefore the Peruvian Indians would surely have seen it, hence the landslide only recently had exposed the opening. He had no clue as to what caused the slide.

Chapter Eight

THE MOUNTAIN CLIMB

José walked away from the slide and Dr. J followed knowing the guide had experience far beyond his. They started their climb, José choosing a route, going away further from their destination. He explained that they would climb a zigzag path to the cave and so long as they moved up the mountain, it made little difference which direction they went. "Of course" José said, "It's much easier traversing rather than climbing straight up," Dr. J realized that he was way out of his league in such matters even though he could not understand why José would start the climb twice as far from the cave as they were from campsite.

After an hour and a half of climbing, Dr. J knew he had lost his bearings on the cave and doubted he could find the campsite on his own. Panicked, he asked José if it was not time to switch back toward the cave. José turned with a smile and seemed ready to speak when a snake slid out of the bushes and crawled straight across the path Dr J wanted to take. José watched the snake disappear in the bushes then looked at Dr. J and said, "We will go this way for a while longer." Dr. J after watching the snake slither across the path that he had suggested they take, then said, "O.K. by me." Dr. J realized that he had panicked for no good reason and that he would have to trust José. Had they taken the route he proposed, they never would have found what José discovered 15 minutes later.

Chapter Nine

STAIRWAY TO THE LIGHT

José slowly traveled toward a line of trees that made dark shade, Dr. J followed about 10 paces behind. José called, "Dr. J, come! See what I have found." Dr. J, with a look of surprise, quickly scrambled up to where José waited. What he saw was undeniable. There he saw a slab rock stairway, albeit barely visible in places that seemed to lead toward the cave. There was scant underbrush because the large trees that seemed to line this stairway protected it with deep shade underneath which nothing could grow. The climb up the stairway seemed so easy. Almost as if the slabs caused those who walked on them to become lighter.

After about four hours into their ascent, José suggested they take a break. Dr. J asked, "José did you know the stairway would be there?´ José replied, "While you checked out the land slide, I observed the mountain and I noticed there was a mysterious line of trees going up toward the land slide's beginning. Dr. J knew only a practiced eye would have even seen this line of trees and he was glad to have José as a guide.

"The best guide in all of Peru," Carlos had said at the coffee house in Lima the day Dr. J arrived. "What kind of trees are these?" Dr. J asked. José smiled and answered, "I do not know but I can tell you they must be thousands of years old." Dr. J pondered that answer for a moment because José had seemed to know the names of all the trees in the Peruvian jungle. Dr. J and José looked out over the valley, now far below in early afternoon sun.

José had packed food for a ten day hike. Finding the stairway put them far ahead of schedule. José built a small fire and made tea. For a long while they sat there in the cool shade sipping their tea and enjoying the incredible view of the valley below. This exciting, perplexing adventure had finally rendered Dr. J speechless as he pondered what might still be ahead. He thought about how much work went into the building of such a stairway since the steps appeared to be carefully laid in place.

José, first to move, broke camp put the teapot and cups back in his back pack and doused the campfire. Dr. J sat motionless throughout this process as the thought finally struck home! It was possible this discovery would be as special as the finding of the ruins of a Mayan city dating back to 100 AD. In fact, this could be even more special. Dr. J suddenly noticed José's readiness and upon standing up was given a pleasant surprise. He did not feel tired at all from the morning's hike because his back and legs felt fresh and not at all sore How could that be? They had walked two or three miles before even starting up the mountain.

Chapter Ten

THE CAVE

The stairway was largely unbroken and allowed them to walk for the next four hours with relative ease. It abruptly stopped at the edge of the landslide. Now they were only 100 yards from the cave, its opening clearly visible from where they stood. They had two hours before sunset and José suggested making camp and going in the next day. Dr. J commented, "Once we get to the cave opening we will have shelter and light so we might as well go ahead up." José saw Dr. J's wisdom in this option and readily agreed.

They made a path to the cave so that it would be easy to return without risking a fall. A single misstep and one would wind up all the way to the bottom. With that understanding, they worked carefully and just at sunset they stepped into the cave. Very quickly it seemed storm clouds had gathered and rain came down moments after they dropped their back packs just inside the entrance. They walked slowly toward the brightly lit room about 30 paces from the entrance.

Dr. J saw a domed room 40 feet in diameter fully lit with no single light source. The emanation seemed to come from everywhere. "After all I've seen I'm not surprised about this lighting phenomenon," Dr. J said to José who mumbled. "I have never seen anything like this, nothing even close!" Dr. J acknowledged José's words, and then turned his attention to the contents of the room. Immediately his focus went to the art work on the walls. The attractively vibrant colors and soft curving shapes did not seem to be pictures but instead geometric forms made up only of circles and the colors and blends of colors struck him as unlike anything he could remember. The walls looked smooth as glass and seamless. He then looked at the pottery shaped in a fashion that one would expect but with unusually fine art work on it. Each pot seemed to radiate a glow of its own. As Dr. J examined one of the pots, he found it to be quite light in weight, the same kind of lightness he felt when climbing the stone steps. Dr. J found this sensation difficult to describe but very real just the same. By this time, the rain pounded, while Dr. J and José stayed high and dry, more than halfway up the side of a mountain in a lighted room with beautiful art work all around. José returned to the entrance to get his pack and prepare some dinner.

Dr. J's eyes covered every centimeter of the room as only a trained archeologist's could. He would study this later but for right now he could smell José's portable kitchen and he felt sure some food would help focus his mind. "I've never seen any thing like this, neither on digs nor in literature. "What have we discovered here?" he wondered while walking back to the entrance where José had set up to cook dinner.

José handed Dr. J a plate and cup of tea, then asked question after question, many of which he could not answer. Dr. J said, "José, I want to thank you, for without you, none of this adventure would have been possible, and I thank you for the food you have prepared on this special night. Gracias Amigo, muchas gracias for all you have done." José now looked at him in a different way as he acknowledged what the doctor had said. He realized that Dr. J spoke from the heart, and felt gratified to hear such words coming from a doctor of Archeology. It made José feel like he shared the same values and thus was on equal footing with the good doctor.

They talked through the evening meal. Dr. J fielded some of José's questions but for most of the questions, he had no answer. One question especially struck Dr. J. was when José asked, "Where does light come from?" He explained, "Ordinarily we think light comes from a source like the sun, a light bulb or a campfire, but inside this room, the light appears to be self-generating. Nothing in science can explain this phenomenon Some unknown factor must have caused the light here and I hope I can find it." But what could that factor be? He noted also that as important as what he found was what he didn't find. He did not find gold, silver or precious gem stones. What kind of civilization would hold pottery and art work above gold and silver?

Soon after the meal and while enjoying a second cup, they both sat with light to their backs and watched the rain as it gentled and softened. The cave light beamed like a search light into the misty rain. They looked at each other and smiled for they knew at this special time it had been a very lucky day. Dr. J thought he might get a good night's sleep and start fresh in the morning. After all, he felt very safe even though he would be sleeping in a cave on the side of a high mountain in the middle of the Peruvian jungle.

Just before he dozed off, he heard José speak, "I feel safe in the cave and you are right about feeling lighter, I feel it too".

When Dr. J awoke the next morning he felt rested like never before. He remembered his dreams for the first time in a long time. All of his dreams were good and they all had something to do with the cave. José, of course, already had some tea and oatmeal prepared. "Buenos Dias," José greeted him now that Dr. J had awakened. "Did you sleep well?" "Yes, I slept better than I can remember. How did you sleep?" Just like you Dr. J. I slept the best too," José replied with a big smile.

Dr. J finished his second cup of Peruvian tea which by this time he preferred to Columbian coffee. As he stood, he felt acutely aware of the apparent lightness of his body and his unusually all over sense of well-being. He walked to the mouth of the cave and surveyed the beauty of the jungle valley below and the distant water falls. He took a few deep slow breaths of air. He had never done that before and did not know what prompted him to do it now but it felt right.

Now, Dr. J with a big smile on his face turned to see José with an equally happy smile on his face. José said, "I sense something very special about this cave and I hope you can find what it is. I know it will be a great help for my friends and their families.""José, I will use all of my scientific training along with a giant computer network back at the University. And you, José, are also part of the resources that I have at hand to solve this mystery."

Dr. J spent the next three hours documenting every square centimeter of the circular room, its art work and the pottery of all sizes located about one meter away from the wall and at intervals about a meter apart. Each seemed placed to allow easy access to all areas of the room. At the center of the room, there was a two meter circular slab of smooth black obsidian that extended 30 centimeters above the floor. Smaller pottery stood around the outside, each piece about one third meter tall and one third meter away from the circular slab of rock. On the circular slab appeared to be writing from some ancient language.

Dr. J could not help but recognize how awake his mind

seemed to be. He noticed his breathing started from his stomach he'd never breathed like that before. He also noted that he no longer felt afraid of anything and as a result felt happy with himself and the world.

Chapter Eleven

THE MESSAGE

Assuming the computer back at the University could decode the language, Dr. J took a picture of the message and sent it via satellite to the University. Just 15 minutes later the freshly decoded message appeared on Dr. J's laptop.

"Welcome to the land of The WILLING. Our civilization lived on this planet 50,000 years ago. For twelve thousand years we lived in peace and harmony until we decided to leave this planet for other adventures. If you too would like to live in peace and harmony in your society, we suggest that everyone who is willing be taught to follow these agreements:

I am WILLING to agree to keep ALOHA in my thoughts, my words and my proactive deeds.

I am WILLING to agree to be responsible for my actions and my Conscious BREATH.

I am WILLING to agree that I am ultimately responsible for everything that happens to ME.

Chapter Twelve

ALOHA

Dr. J knew right away that he could not understand the first agreement without complete information about ALOHA. Dr. J called the language Dept. at the University of Hawaii in Hilo on the Island of Hawaii. Soon he spoke with Dr. Sam Kaluna, a Hawaiian language teacher.

"Hello, Dr. Kaluna, I'm Dr Jayson, I am calling from the Peruvian Jungle where I have made a discovery. I find that I need your help." "What can I do for you Doctor?" answered Dr. Kaluna. Dr..J says, "I need to know about the word ALOHA."

When Dr. Kaluna heard that, he laughed and said, "We are still working on that one in the state of Hawaii, but Hawaiians have always known its meaning. Aloha is truth, compassion, and the intelligence that guides us. "Aloha" is the Hawaiian mode of understanding relationships characterized by openness, connection, humility, and love. Aloha is feeling the other, seeing the other, and caring for them." Dr. Kaluna continued, "Aloha is what makes Hawai`i so beloved. It is the way of the breath that makes us smart. Here is our core animating principle:

ALOHA

"Before you can have Aloha you must do Aloha

Before you do Aloha you must be Aloha.

Before you be Aloha you must choose Aloha.

Before you choose Aloha you must have the power to choose Aloha.

And before you have the power to choose Aloha.

You must have consciousness of the source."

A = Ai'Akabai = Being kind and tender!

L = Lokabi = Being helpful, cooperative and harmonious!

O = Olu'Olu = Being agreeable and pleasant!

H = Ha'aba'a = Being humble and modest!

A = Abunui = Being pleasant and persevering!

By: Piloli Paki

Aloha is not distinct to Hawai`i, though it is bountiful here. Hawaiians always have demonstrated it to others who moved here from far off lands. The lessons have not always met with success, but there is a general understanding of the concept of ALOHA here in Hawaii." He continued, "In my opinion aloha has made Hawaii a special place to live or visit, because of its acceptance and acknowledgement that all people are capable of receiving and giving love every day with every breath, in every situation." Dr. Kaluna continued by saying that ALO (ALO-HA) means 'in the presence of' and HA means 'breath'. Dr. Kaluna went on to say that while Aloha means Hello or Goodbye, it also, and more importantly, means to seek an intimate relationship; or it can also mean to look for the best in any person that you might meet." Dr. Kaluna went on to say that when a Hawaiian is full of ALOHA, he has special mana or the power to do good, because he is full of love energy. Then Dr Kaluna quoting William James, said "The greatest discovery in my generation is that human beings can alter their lives by altering their attitudes of mind," and that is the insight of Aloha. Aloha is a concept, where upon even seeing another person, one might summon up good feelings toward that person.

When Dr. J hung up the phone, he knew he'd discovered something big. No wonder that the computer could not find another word for Aloha. Aloha is the gift from Hawaii for the world. Bigger than anything man could invent. The concept shows how to live in peace and harmony. This was the ultimate secret to the universe, the answer to every man's desire. So simple it was in front of our faces and we could not see it. Somehow, all known past civilizations seemed to think that gold was important. So important became gold that the entire concept of ALOHA was lost in most commercial centers. The truth suddenly flashed in Dr. J's mind. Twelve thousand years of practicing Aloha created energy so great it became light. This is the energy that the Hawaiians referred to as mana. The artwork and pottery in the cave are the source of the light. To prove his theory, that night Dr. J took a piece of pottery out in the dark and sure enough, it gave off light of its own. The loving energy put into making the pottery or painting the dome gave it light of it's own. It is amazing what the mind is capable of when directed with ALOHA.

Now that I understand the concept of ALOHA, I also understand that if people were taught the secret of ALOHA, then they could teach it to their children. If they are WILLING they will be HAPPY. And after all is said and done, people just want, more than anything else in the world to be HAPPY. Forget about getting rich. If you earn enough money to pay all your obligations then you can be HAPPY and productive and creative to do anything. But first the agreements must be understood and also the concepts of ALOHA.

"Make ALOHA your gold" should be the bumper sticker of the future thought Dr. J. He found himself walking around and around the ALOHA dome breathing and smiling. Then he remembered the old woman who had said, "Your family's happiness will be found in the side of a mountain." Dr. J recalled his reaction when he first heard the woman say that. He laughed out loud at the thought. What he took as absurd turned out to be the most exiting information yet in the history of humankind.

Dr. J spent another three days documenting all aspects of the domed room inside the cave. He wrote notes on everything that happened in his journal as trained scientists are taught. Now was the time to return and report to the Peruvian University Department of Archeology in Lima.

Chapter Thirteen

THE WALK BACK

Journal Report:

March 3: Late afternoon:

On the hike back to Lima, we stopped at the village of the bone throwing Wise Woman. José told the village people that what we had found in the side of a mountain matched what the Bone Thrower had said. Dr. J in his best Spanish told the villagers about ALOHA and the Three Agreements. Then Dr. J set the piece of pottery from the cave down and took off the covering; it, of course, gave off light of its own. This sparked a celebration that evening including music, food and a lot of happy people dancing and laughing and having a very good time. The old woman said the Agreements will help the village people just as it will help people every where, with this ancient new way of thinking and acting. The woman said she was happy to have met me! I smiled and hugged her. Then I told her that I was most happy to have met her and I thanked her for helping me find what I wanted for my family.

By nightfall after a wonderful meal, the village was still awake and lively. The village chief commented that some grudges that had lasted for many years, dissolved in the blink of an eye once they heard and understood the Agreements. After all, we all want to be HAPPY above all else and now we know how to make it happen.

March 4-5: side trip

We took a long side trip so José could show me some sites in the area that he loved. He brought me to a giant waterfall with crystal-clear swimming pools and led me down a path to the multi-colored birds gathering place. No camera could have done justice to their striking colors, brilliant, iridescent and stunningly attractive. The best guide in Peru....True.

March 6 :Lima

The Archeology Department at the University in Lima made a holiday out of our return. It seemed everyone somehow knew all of the Agreements before I could tell them. When someone asked the Chancellor of the University how this happened, he said only that in Peru good news travels fast and really good news travels telepathically. I cannot argue because all I have to say about the discovery, they already know.

I asked the chancellor about the artwork. The chancellor went on to say, "The geometric form in the dome and on the pottery is called "The Flower of Life" which is the base of something called sacred geometry. It is a very new science. Basically "The Flower of Life" symbol has the ability to demonstrate how all things come from one source and are ultimately and permanently woven together. The symbol can be used as a metaphor to illustrate the connective- ness of all life and spirit within the universe".

Everyone seems quite happy with reported facts, almost as if they had been there in the circular room with me as I documented the findings. José just winked and said "ALOHA, Dr. J, Aloha conquers all including time and space." I agreed with that assessment. There in the Archeology Department's lecture hall, I placed the pottery container with "The Flower of Life" symbol on the desk. The entire audience hushed as I .switched off the lights. The thirty-centimeter-tall container radiated a beautiful soft light, easily enough light from which to read. Everyone cheered then laughed and cried tears of joy then cheered again.

March 7-8:

My two new friends from the coffee shop invited me to an evening. I told the families about my findings and the Agreements. Already, a new day had dawned. Each family quickly got the idea and almost from the very beginning started laughing and talking about old attitudes that no longer apply.

March 9

Contacted the Supervisor of the dig site and explained my desire to return home and share my findings with my family and the University of Nebraska Archeology Department. The supervisor enjoyed talking with me and said he more than understood the situation and supported my request. He also stated that a full complement of scientists had arrived at the cave. They confirmed that a 50,000 year old time bomb had exploded and caused the land slide that exposed the cave opening. He commented that 75% of the pottery (about 100 pieces) was being sent to great universities and museums around the world along with full information about the Three Agreements. He felt we had discovered the most powerful force in the universe. This is the key to end all wars and all domestic violence. Also the most important find ever because after all it shows us how to be HAPPY by knowing how to conduct our lives.

Chapter Fourteen

FINE FOOD AND PICNICS

March 10

I invited José out to eat in a really fine restaurant. After we had ordered and while sipping on some light wine, José mentioned that even though he had lost most of the lightness, he felt in the cave, he could get a lot of it back just by breathing and keeping ALOHA (positive) thoughts alive in his mind. Also just by breathing and being aware of breathing and keeping ALOHA with no thought was a wonderful way to be. That is: if nothing is happening at the moment- no need to be using precious thought energy for no purpose. Just keep ALOHA and follow the Breath.

I thanked José for putting it in words like I never have. By showing me this insight, he had opened up a whole new world for me. My family will be grateful once they hear about the way of THE WILLING

March 11

Spent the afternoon attending a picnic with the families of the two men I met at the coffee shop on my first day in Lima and three other families. After a wonderful home-cooked meal consisting of locally grown fruits and vegetables, I spoke to the families about our trip and the discovery of the cave and its contents. As soon as I told them of the Agreements, each individual in attendance appeared to become more relaxed and less competitive. It seemed the children picked up first on the Agreements and the BREATH and the ALOHA light. They saw this could lead one directly to a HAPPY mind. We heard one child say, 'Now that I understand about the BREATH and keeping ALOHA in thought, I can see how it would be easy to be HAPPY, just by following the Agreements. Keep breathing and live ALOHA." The adults sat there on the picnic grass silent and dumbstruck as they processed the Agreements themselves and how they might fit into their own lives. Also they started breathing deep and slow from their stomach and then breathing slowly out.

I reminded the families about the differences and the ways they might have thought in the past. I told them a complete change will not happen overnight but we can all attain happiness by sticking to the Agreements and paying attention to the even flowing BREATH OF ALOHA. By the picnic's end, the mothers praised this new way of thinking. They laughed happily. They danced around,, first with their children and then with their husbands. They even got me out there dancing too and of course I had a wonderful time dancing in the light of ALOHA. We all had a great day. The children all played well together with no fighting.

With the setting of the sun we said our goodbyes and hugged each other with tears of joy. We promised to keep in touch and to exchange any new insights into the use of the Agreements and the ALOHA BREATH.

ALOHA is looking for an intimate relationship with one another and always looking for the best in each other.

March 12-13

Spent time relaxing and sightseeing around Lima I sent word back to my family that I would be home by the 16th and to please pick me up at the airport. I also e-mailed the University of Nebraska. The answer from my family sounded optimistic but the University withheld enthusiasm. They wanted to scientifically test the pottery to determine the source of the pottery's ability to give off light. They said the scientific study might take a couple of years. They seemed to be no scientific interest in the Three Agreements. It came to me clear right then that introducing the Agreements can only happen one person at a time because science knows nothing about the power of ALOHA.

Chapter Fifteen

TICKET TALK

March 14-15

Traveled back to San Francisco and then eventually to Lincoln, Nebraska.

On the plane to Lincoln I talked to my fellow passengers. I listened to what they had to say. All too often there seemed to be a false tone coming from an oxygen deprived brain due to too much shallow breathing and breath-holding. Many stated that their family problems almost overwhelmed them. I heard stories of children acting in a non-ALOHA manner towards their mother. A mother, who has so much to do, cannot give her children the quality time and attention they deserve. Their children all seem to watch TV in excess and afterwards have no energy or creativity to do anything worthwhile, least of all their homework. And they don't even read books anymore.

One parent commented that after her son's grades hit bottom, she limited him and herself to one hour of TV on week nights. She reported that her son's grades went up. She found the TV commercials horrifying and very chaotic for young minds to absorb. She switched to PBS during the 3-5 min adds. She used an egg timer to allow conversation in the quiet room during the break. She also says that she uses the mute button when a commercial comes on if she does not switch channels to PBS.

I noted the wisdom and strength of that mother and decided to do the same with my family. I saw in these folks no fault because they believe they think normal thoughts, and have simple normal conversations. Until they learn about the agreements and the BREATH of ALOHA, they just don't know how to act or talk or teach about happiness. We live in a world where war an fear are promoted. These promoters dutifully tell us what to fear next, or buy next.

The plane ride home made stops in San Francisco, Denver, Kansas City and finally in Lincoln. So I had plenty of opportunities for discussion. Of course, I could add that not all the conversations were negative, not even the bad ones were bad all the time. I realized that before I became WILLING, I would have never noticed the differences.

Chapter Sixteen

HOME AGAIN

March 16

Dr. J arrived at 9:05 and had only a short wait before Sue arrived after having taken Patty and Phillip to school. They hugged and kissed happily before picking up the luggage. Sue noticed Dr. J radiated a beaming smile and felt it was making her feel good too. When Dr. J asked her about home life the past month, Sue admitted that the days ranged from not very good to mostly normal and sometimes great. "We've been eating well, have clean clothes, we're in good health and all the bills are paid so we are in pretty good shape despite the fact that we missed your help to get past the dramas."

Sue and Dr. J spent some quiet time together before Patty and Phillip got home from school. Dr. J also took a long hot bath and a nap, and then spent some time in the garden with the flowers and of course with Sphinx, that very friendly lazy orange cat.

Phillip and Patty greeted their dad with happy smiles and

hugs. Dr. J and the children had some quality time together as Dr. J learned about the school adventures and life at home the last month. He then briefly told them about this trip and promised to tell more at the dinner table.

After talking and laughing and telling stories time just seemed to fly. So an hour and a half later, Patty and Phillip set out to do their evening chores and take care of homework before dinner. Dr. J unpacked his clothes along with the presents that he brought back for his family then hurried down stairs to help Sue with the evening meal.

Chapter Seventeen

DINNER TALK

Tonight would be a tofu stir-fry with fresh garden salad and a special sauce that Dr. J had learned from the mothers at the family picnic in Peru.

Phillip had already set the table for the meal as part of his evening chores. Sue called for Patty, who had cleaned her room and the bathroom before coming down for the evening meal. The family gathered at the table and gave thanks to the Great Spirit for the bountiful meal. Dr. J said 'There is something I did not tell you yet, because I wanted you all to hear it together.'

As everyone began to eat, Dr. J told the family of the Three Agreements, the BREATHS and his understanding of ALOHA.

First of all, "ALO" means "in the presence of" and "HA" means BREATH. So I understand it to say: to keep ALOHA, you must follow the BREATH by breathing slowly and deep from the stomach. ALOHA means to seek an intimate relationship with self and others. ALOHA means to truly accept and love yourself. ALOHA also means to look for the best in others, always look for and find the ALOHA, even if it is not quite there or not there at all, you still give ALOHA. A non-ALOHA person will probably get it the next time and give ALOHA back to you. Don't worry if it doesn't happen because worry is a non-ALOHA thought pattern. Just keep breathing and give ALOHA and give ALOHA and keep giving ALOHA to everyone every day. It's easy to do and so much fun especially when you get ALOHA back. Change the way you look at the world and the world will change right before your eyes. Now that I have helped you with your understanding of ALOHA, let me tell you about the BREATH.

We must make it a constant practice to breathe in slowly and fully and exhale slowly and fully. We should fill the stomach and then the lungs all the way to the top. As you inhale, think about beauty or think about freedom or think about ALOHA. As you exhale think gratitude, thankfulness and humility for the wonderful gift of a life in which to practice ALOHA. It is very important to keep the breath connected under normal activites, slow and deep- no shallow or freeze breaths please.

First thing, each morning, to kick start the body, blood, heart and mind it would be wise to wash your face with cold water, then as soon as possible do the WILLING BREATH. Sit with a straight back, touching the tips of the first two fingers and thumb of the left hand, then loop first two fingers and thumb of the right hand. If comfortable close eyes and moment by moment breathing the WILLING BREATH that consists of seven rounds of seven and the seven colors, of the rainbow ,one at a time, first breath—

red, second-orange, third-yellow the onward with green, blue, indigo [purple], and violet. A good way to remember the colors in sequence is: Roy G Biv. See each color in your mind's eye as you breathe in from your abdomen on seven counts (Always thru the nose)

Hold for a count of seven.

Exhale for a count of seven.

Hold for a count of seven then start over making seven rounds in all

This exercise will make you wide awake and relaxed. An excellent way to start a HAPPY day. It is also a good way to end the day by doing seven rounds of wonderful life giving breaths to insure a good night's sleep.

This next breath is simple way to build immunity and eliminate disease. "DEEP BREATHING EXERCISES THAT OXYYGENATE THE CELLS ARE MOST EFFECTIVE WAY TO LIVE WITHOUT DISEASE IN YOIR LIFE" Dr. Otto Warburg, two time Nobel Prize winner.

It's true-the medical evidence is overwhelming! Supplying your body with more oxygen will burn fat, slow down the aging process, prevent heart disease and magnify your energy. You will drastically lower your risk of disease. And you will be more successful in every area of your life when your body is functioning properly.

Truth is, there is no easier way to achieve great health than to do aerobic breathing once a day and drink plenty of the purist water you find, spring water, if possible. It's simple.

You can do it in the car on the way to school or work, during a commercial break while watching TV-any time or place that is convenient to you. This is a basic maintenance program to keep your body firm, healthy and attractive. Here is how you do it: First, imagine that you are blowing through a straw. Blow as long as you comfortably can. Then Sniff up as if you had a runny nose and hold it to a count of three. Now with your mouth closed, force the air through your mouth then let it drop open. Make your mouth as big as you can while you force all of the air out of your lungs. The sound you should hear will sound like air escaping a balloon Then hold your breath and contract all the muscles in your stomach and diaphragm, lifting them up toward your throat. Hold this position as long as you can. Now take a normal breath and start again.

The secret is to do this at least ten times in a row every day. You'll flood you cells with life-giving oxygen!

For the normal deep breath here is a Hawaiian technique called pikopiko, because piko means both the crown of the head and the navel. This information is brought forward by Serge Kahili King, a world famous teacher of ALOHA PHILLOSPHY, and made available in over 22 languages world wide. Here is how to Enhance your power to bless by increasing your personal energy. It is a simple way of breathing that is also used for grounding, centering, meditation and healing. It requires no special place or posture and may be done while moving or still, busy or resting, with eyes open or closed. In Hawaiian the technique is called pikopiko because piko means both the crown of the head and the navel. The technique: Become aware of your normal breathing {it might change on its own just because of your awareness, but that's okay] Next locate the crown of your head and your navel by awareness and/or touch. Now as you inhale put your attention on the crown of your head; and as you exhale put your attention on your navel. Keep breathing this way as long as you like. Very good! When you feel relaxed, centered, and/or energized, begin imagining that you are surrounded with an invisible cloud of light or electro-magnetic field, and that your breathing increases the energy of this cloud or field. As you bless, imagine that the object of your blessing is surrounded with some of the same energy that surrounds you.

Here is another exercise you can use during the day as you stay close to the BREATH. This breath is for stressful situations or when you wish to calm your mind. Please know that the '8-4' breath can be further enhanced by counting the heart beats. Or you can count syllables instead of numbers as 1-2-3--8 and 1-2-3-4 and so forth. You might say to yourself, "Self be loved, self live ALOHA" as you slowly inhale. Then say "I am HAP-PY", for the four counts, then say (during the slow exhalation) "Self be loved, self live ALOHA" that's good for eight counts, then while holding for four counts, say "I am happy". As you do this exercise listen to the meaning of the words. If you desire happiness, you will embrace these words (concepts) so that they become part of you always.

"Faith is an invisible magnet, and attracts to itself whatever it fervently desires and calmly and persistently expects."

I know all of these different 'breaths' might be over whelming at first so re-read them and do the ones you feel comfortable with. You can always add new ones later.

Chapter Eighteen

TALK OF LOVE

Before I say anything about thoughts in general, let me say that it is very important to love oneself, and be satisfied with self.

To be HAPPY, one must know self-love and be comfortable with self at all times. Happiness is based on self-acceptances. Embrace your self thus it will be natural to embrace and love others. When we understand that love is the most powerful force in the universe and it also is the force that holds families together thru all sorts of trials. So do not be discouraged about any aspect of yourself that you feel is less than perfect because when you keep giving yourself love (ALOHA) you shine the light for all to see. So ALOHA yourself in all ways first, then every thing else around you is easy and natural to ALOHA. Remember you are perfect just like you are so just love-self, love-others and be HAPPY because of it. Also concerning self-love, if we understand that we are part of Mother Earth, then our self-love becomes love for all creatures and plants and all that we know that is.

Love is the binder that keeps life here together.

Without love we are nothing.

Chapter Nineteen

THE EGG AND I

Now I am going to ask that you use your imagination. Your imagination is one of the most wonderful gifts that you possess. You can use your imagination to understand concepts and ideas that cannot be seen. So here goes: See yourself as a freshly laid goose egg. See the outer shell as your protection skeleton. See the clear liquid inside which we will call the "ocean of emotion." See the golden ball inside the ocean as being you, the TRUTH. You can access any part of the "ocean" that you want at any time. It is important to note that the "ocean" contains ALOHA life forms and also non-ALOHA life forms. The non-ALOHA life forms are (in part as there are many) fear, anger, sadness, regret, hatred, jealousy, judgment, selfishness and "stuck in the mud"-laziness. If you allow or access any of these non-ALOHA life forms into you, the golden ball, they will cause your BREATH to become very shallow or even stop while the non-ALOHA emotion takes over your life until it runs out of steam. When a non-ALOHA emotion is in control you are no longer the TRUTH and too many times you have to pay dearly for the consequences of the non-ALOHA energy form that rampages through your mind and body.

So with practice and careful observation, you can tell where the non- ALOHA energy 'form' is entering your golden orb and then you can seal it off with the full slow stomach expanding BREATH. Do not let emotions such as fear or anger gain control of you. You have a clear choice and that choice is to be part of ALOHA-energy (harmony, joy, unity, wholeness and of course, happiness).

No matter the situation the choice is always the same. You are part of Aloha—nothing else matters.

Non-ALOHA energy can cause the tones of your words to be harmful for others to hear and your non-ALOHA actions may also be hurtful to others and only you are responsible, therefore when this non-ALOHA energy goes out in the universe and reverberates back upon you, some say it will be ten times as bad.

You can best know by practice and observation of yourself how to spot a change of your attitude. If you really desire happiness it is good to use this example of the goose egg to better understand emotions and where they might come from and how you can protect yourself. That is, when we learn to feel our body when we breathe, we are also learning to use no-thought for greater self-awareness.

The more self-aware we are the easier it is to spot an upset wanting to erupt. We learn to remain calm, breathe, then move forward with ALOHA even though it may be difficult. You must be willing to let the "storm pass" while you use courage to stay calm, seek harmony, and speak the TRUTH.

Most often non-ALOHA emotions are brought on by lack of clear communication or by someone attempting to impose their will in a situation. Usually when someone feels the need to impose their will it generates non-ALOHA feelings in return.

Communication will always do just fine with courage, the BREATH and ALOHA. With the intent to seek harmony, you can be HAPPY anytime and in any situation. By keeping the beast (non-ALOHA emotions) locked away, or breathed away before it can gain power, we can think, speak and act in ways that will inspire others to also be WILL-ING to be HAPPY.

Non-ALOHA emotions never solve problems, they only postpone them. ALOHA emotions are very good for solving problems.

A happiness tool is aromatherapy. The various aromas can spark ALOHA emotions. Go to the health food store and take a free sniff. Look for grapefruit. You will surprised what emotion it will spark. Aromas are like magical spaceships, they can transport you. Try them, you'll see for yourself. Also a sniff of peperment right before a test will help you to concentrate.

We are all in pursuit of happiness. Any tools we can use to help out in our pursuit are welcome.

Chapter Twenty

THE FIRST AGREEMENT: ALOHA IN THOUGHT

During the day when your mind wanders into non-ALOHA territory, just go to a few 8-4-8-4 slow breaths and think of the beauty of nature or how it feels to be happy. Think of how you can and must love yourself because the prize is happiness and you will be back where you want to be, and you did not mess up your future by thinking non-ALOHA thoughts. You may need to do this often at first so be confident if you desire to be happy. Remember: "We will be tomorrow what we think about today, so therefore it's smart to always keep the BREATH and ALOHA thoughts very close to the heart of happiness. It is said "Thoughts are things." It is true what we think about is what we are about. By keeping ALOHA in our thoughts we hook up with our spirit. Spirit, the source of all creativity, wants us to be HAPPY so if we want the same we must do our part. Good people keep good thoughts. HAPPY people keep HAPPY thoughts. It's easy to be good when you are HAPPY. It is important to always keep the breath connected, and thus connected with spirit, the source of all things good.

When something happens that upsets you, it is best to be smart and use your courage to breath and send your awareness to your stomach. The non-aloha energy will vanish if you do not give it energy by thinking about it. Also it should be said here that we should avoid stimulants such as sugary and/or salty foods and drinks. Soon you will lose the taste for such when you realize that the unnecessarily stimulated body does not allow for a calm mind. The calm mind keeps the door open for your greatest gift; the creative spirit.

Chapter Twenty One

FIRST AGREEMENT:
ALOHA IN WORD AND DEED

Never underestimate the power of the spoken word. Since words have power, we should use ALOHA words, paying special attention to the tones. People feel and understand the tone of words even if they do not understand the words. Keep the tones inside the light of ALOHA and speak from the stomach (diaphragm) When you hear your voice pitch go too high you are probably speaking from emotional imbalance and not from a relaxed ALOHA state. Stop talking and gather your breath before you speak again.

You will find no joy in non-ALOHA thoughts and actions. So just leave them alone. If someone in your presence speaks non-ALOHA words, do not encourage them. You can either say nothing or better yet change the subject to one that lives with ALOHA and remember to keep the life giving, fun loving ALOHA BREATH flowing deep and slow. Soon you will be laughing and having fun with the "Light of ALOHA" in the presence of you and your friends again. In other words keep the ball up always and if the ball does go down wait for the "bounce" then keep the ball UP.

Sue sat there at the table with a soft comfortable look on her face. She smiled serenely as she processed the information. As Dr. J presented the information she observed the way Patty and Phillip received the message. Patty and Phillip were already practicing the BREATH so when Dr. J stopped talking and took a bite of food, Patty said she realized that her stomach had been too tight. "Now my stomach is relaxed for the next breath. My mind seems free of troubles." Phillip said, "I can see the difference already in my attitude." Softly Phillip said "I am already more relaxed."

Dr. J went on "To keep ALOHA in your proactive deeds is also part of the First Agreement. Proactive means to do things with forethought. When you act after an action it is called reaction. It is more difficult to be in control when one reacts to (say) an accident or something unforeseen. If you are calm and balanced you will think and breath before you decide on how to respond. You might need to know that when you do your chores without being told (for example), that is proactive. Therefore always keep ALOHA when you do your chores, homework or anything that requires effort. It is very good for you and everyone around when you keep ALOHA in your proactive deeds. It is good to be proactive. That means to do things that will help yourself or others. Act always with the BREATH and profound ALOHA in your being. You will be connected to the source and it will be with you as you stand in the *light*.

Chapter Twenty Two

THE SECOND AGREEMENT

Now let's look at the second agreement which says, "I am willing to be responsible for my actions and my BREATH."

Remember this: "to accomplish GREAT THINGS, we must not only act, but also dream; not only plan, but also believe." "We should be taught not to wait for **inspiration** to start a thing. Action always generates inspiration.' "Whatever you do make sure you do it with ALOHA. If your action goes in an unintended way, you must be responsible for the action. That is: You must fix, pay for, apologize and admit to any possible misdeeds. Admit the truth. After all, that is what we want to be all of the time. We must be willing to use our courage. Tell the truth always. Tell the TRUTH and one will live forever. Also, who is responsible for your breathing process? Who will breathe in the oxygen for your mind and body and happiness if not you?" So you must be responsible at all times because your happiness depends on it."

Your facial expression is an action to which we should give

some attention. If you are smiling, other people will usually smile back. That's a good thing." So Smile a lot, not a false plastic smile but a smile that reflects how you feel inside. Always keep the Face of ALOHA handy for all to see. A smile is an inexpensive way to improve your good looks.

"Remember when you tell the TRUTH you say who you are. Sometimes it takes courage to say who you are."

Chapter Twenty Three

THE THIRD AGREEMENT

And lastly, the Third Agreement says: "I am willing to agree that I an ultimately responsible for everything that happens to me."

"This is a tough one because it does not allow for the blame game."

If you are walking in the woods and you trip, fall and jam the thumb on your writing hand, who can you blame? Can you blame the rock that you tripped on? Well, the same goes for everything that happens to you. You must understand that past non-ALOHA thoughts or actions may have had something to do with your present situation. Stay out of the "blame someone or something game". It is not the road to happiness. Just deal with it and keep the BREATH, ALOHA and the Agreements close to your heart. All good things are on the way and happiness will live in your world."

Dr. J realized that he had talked away the dinner hour and had eaten very little. So he stopped talking and turned his attention to the food on his plate.

Chapter Twenty Four

JOY TO THE WORLD

By this time every one at the table had finished. Patty and Phillip sat quietly, obviously focusing on their breath. Sue was beaming. She said, "This is very important information. Everyone on the planet needs to know this. The people need to know this to regain control of big business, the military, the corporate monopoly and a government that seems to think only of the top two thirds of the population are necessary to make the economy grow. Do the math. Two thirds of the 300 million population in the United States is 200 million. That leaves 100 million people who may have lost their home and now live in poverty because of economic disenfranchisement. One hundred million is more than the population of California, Texas and Oklahoma combined. With this information, we, the people, can breathe all the way to the top and regain control. Again, first we must teach the children. The children then teach other children and the children then also teach their parents. But first we must teach the children to stand in the Light of ALOHA."

Then Dr. J said "The children can then support each other by forming "WILLING groups." These WILLING groups can do projects, write and read poetry, write and produce their own music or plays, discuss books, play games, exercise and eat snacks. Snacks should not only be good but nourishing too. You can have so much fun with friends who are WILLING. Stay away often from salty chips, white sugar cookies and sodas, all of which are non-ALOHA for the mind and the body. Also I might add the new guidelines for the T.V.. Only one hour every night except Saturday and most important we WILL mute out all commercials. Commercials are for the most part non-ALOHA in spirit thus contain nothing to enhance your happiness. So just look away from the T.V. while a muted commercial is on. Check where your energy is and if necessary recharge your battery with the conscious breath.

Chapter Twenty Five

PATTY SPEAKS OUT

"Let's face it," Patty said, "Most shows on TV are simply not cool anymore. They rely too much on visual violence and after all that is a long way from ALOHA? Why invite violent words and mean spirited tones into the living room on purpose. How does that kind of energy nurture me? Energy that promotes fear will not and cannot make me feel love. It is just the opposite. So tell me what is going on? TV business people want us to watch the show and buy what the commercial says. They do not necessarily care for us to be HAPPY. Our happiness does not appear to be of their concern.

"Since most people can be tricked into stop breath where the non-ALOHA emotions take over the TV, people make their money by selling fear to people who do not understand how to breathe. Fear is quite unattractive and useless to someone full of ALOHA," says Dr. J.

After a month of the family practicing the WILLING BREATH, it seemed like the old disharmony game was over for good. Now it is live ALOHA and help support others. (Support others to mute the commercials too.) When they slip and drop the "ball" simply by reminding them to breathe a few slow ones and do it with ALOHA. If they are WILLING, they will be right back in the comfort zone.

In only this few weeks, their home almost ran itself. There was much laughter and harmony in the home. A few slip-ups happened, of course, such as getting caught in a situation which seemed at the moment more important than the Agreements, but always when the BREATH returns slow and full the TRUTH returns to the present for all to see.

Dr. J said, "It's important to note that when one gets close to anger or any non-ALOHA emotion, the breathing stops. That emotion then is in control of you. To be HAPPY it is good to learn to control your emotions, and you will, as you become comfortable with the TRUTH. There are certain cultures such as the Hawaiian Culture which says we have two brains: one in our head and the other over our stomach. You will connect with your stomach brain the more you put your mind into a relaxed stomach. The good thing about the stomach brain is the opportunity for no-thought to happen. It takes energy to keep the head brain thinking. So it is a good idea not to over heat the brain but instead, let it relax. You can go to the stomach brain where no-thought is happening as you are pumping fresh oxygen to "both" brains and your body. When 'no thought' is going on the spirit of creativity may take the opportunity come forth. We are at our best when we speak our TRUTH through our creativity.

Chapter Twenty Six

GUT FEELING

Speaking of the stomach brain, have you ever heard someone talk about intuition? When you learn to listen to your intuition it will always guide you to the correct solution. Only when we shut down all the clatter in our head brain do we allow for the TRUTH to come through. Have you ever heard the term "gut feeling? Science does not know where intuition comes from. Women have long been accepted as having intuition. Men always say I have a "gut feeling." When you have a relationship with your full slow breath, you become very familiar with your stomach brain, which will open the door for your newly found gift of intuition, or "gut" feeling.

If we stop breathing, we then set our self up for non-ALOHA thought patterns which in time open the door for the emotional field to rush in. The worst of the emotional field is fear. Fear controls most people some of the time and some people all of the time. When fear comes knocking on your door just expand your stomach and breath right through it. (When the stomach gets tight, the emotions can gain control.) You will be amazed at what a laughable emotion fear is when you breathe correctly.

Most fear comes from the unknown. When you learn about the unknown as best you can, then you know more and you fear less or best yet you have no fear at all when ALOHA stands strong in its place."

Sue walked over and gave Dr. J a loving hug and kiss, then says "You are wise, Dr. J, and I am going to remember and use this information. Patty and Phillip were laughing, jumping and skipping around the room and thus made a personal discovery: they now know the importance of exercise because it keeps the blood pumping to the brain.

Then Sue said I have written a poem called "Ergo Life" that I'd like to read at this time.". Then Sue begins to read:

ERGO LIFE

Please refrain

From placing blame

And do the same

When you could complain

It serves no one

To bring life down

For if that goes out

It comes back around

With way too many

Discordant sounds.

Now if you have an axe to grind
Try looking within- you'll do just fine
Fill your lungs both full and slow
With equal pace, you let air go.

Both full and slow, both in and out
You become aware of what joy's about
The joys of life, to say seems fair
Are best experienced with ample air.

The shallow breath that's froze for a time
Has forgotten the rhythm that makes life rhyme
Our wireless connection to life is quite clear
But when short on air, it becomes hard to hear

So when the brain
Becomes way overheated
Go to the breath
For the calm that is needed

As it is above so it is below
A happy life thrives on the full even flow"

Chapter Twenty Seven

NON VIOLENT SOLUTIONS

The following Monday at school, the English teacher, Mrs. Janet Lui gave a writing assignment to be due on Friday. The theme of the assignment was "Non Violent Solutions." That day after school as Phillip walked home, two boys from his English class jumped out of the bushes and stopped him. They wanted Phillip to write themes for them or they would beat him up and steal his books. Phillip caught himself holding his breath with fear, so he relaxed his stomach and took a good deep slow breath and his fear disappeared. He decided to do what the boys wanted. So he agreed to write the papers for the two boys. Then Phillip stopped by the library. When Phillip came home after school with a load of books in his back pack, Sue looked at the stack of books and wondered what kind of research Phillip was doing, but thought the dinner table would be a better place to find out about the mysterious stack of books.

Meanwhile Patty was not quite herself. She had in the last month changed her attitude so much that for the first time she seemed to need a heart to hear. Sue met with Patty in her room.

Patty said, "Mom, something happened at school today and I need to tell you about it." She then told of her classmate's behavior at school when they heard of the cancellation of the museum field trip.

It so happened that the bus driver's wife became bed-ridden. She could not take care of her two small children, one of whom was still in diapers. She hurt her shoulder and badly sprained her ankle when the neighbor's dog chased a cat under the car she was washing. In an effort to get out of the way, she got tangled in the garden hose and thus twisted her ankle and hurt her shoulder when she fell on the cement.

The teacher explained the incident to the students who for the most part expressed sympathy until they heard the part about the field trip being cancelled.

Some of the students argued that they should not have to stay in school and miss the museum trip because of a dog. The teacher felt the anger in the argumentative students. Patty felt that anger multiplied by others who bought into the "Blame someone else game." Class mates, who Patty had introduced into the WILLING remained calm.

Patty remembered to take a step back and breathe four or five slow breaths and after she did, she checked to see how she felt, and how she felt about the dog and the garden hose and the accidental injury of the bus driver's wife. Patty realized that she had no part in the class's anger. She committed herself to finding an ALOHA solution.

A lull finally came in the exchange between the angry students and the teacher and Patty offered a possible remedy to the problem. She told the class that she would ask her Mom to go to the bus driver's home and stay with the laid-up mother and care for the two young children for the morning and maybe someone else could figure out how to help in the afternoon. One angry girl asked, "Why can't your mom stay for the whole day?" Patty knew then that things were out of control. The girl assumed Patty's Mom already agreed to help in the morning. Patty said, "I only wanted to go to the museum just like everyone else. I need to ask my mom first. We are all getting ahead of ourselves. Let me ask Mom tonight and I'll report back in the morning."

Patty spoke with her WILLING classmates during recess. One of the classmates said, "What a good lesson you taught us in class. Patty, you were so calm when you spoke to the class, you made me more willing than ever to pay attention to my breath. Before you spoke in class, I admit that I was holding my breath with fear. When I took some deep slow ones, my fear disappeared like fudge on a plate." Everyone laughed then got ready to go to the next class.

So while Patty and her mom were still alone, Patty said to her mother, "I hope you can help out, mom, but I know it would be asking quite a lot of you, especially since I didn't ask you first."

Sue had listened very carefully to all that Patty had to say. She smiled warmly and said, "Patty, you are doing so well. You willingly share with others the joy of harmony and unity. But, Patty, there is a problem with your idea. I have an appointment with the dentist at 11:00 Friday morning so I cannot help out as you desired. Perhaps I could help out in the afternoon." Just then the phone rang. Sue excused herself and went to answer it while Patty put away her freshly washed clothes, her mom had stacked on her bed that day. Patty felt good about her clearer understanding of how one is supposed to keep control of the emotions. Patty remembered the non-ALOHA tones she heard from the classmates who had let anger take control. As Patty put away her clothes, she took the opportunity to breathe slowly and nurturing. She felt very peaceful and comfortable as she folded the clothes and carefully positioned them in their place.

Dr. J helped out in the kitchen . Phillip had come to enjoy setting the table and placing the napkins just so. Phillip always did a great job since he became "WILLING". Later Patty and Phillip did the evening meal dishes, and wiped the kitchen counter down as part of their new chores. Chores, after all, are nothing more than daily opportunities to show Aloha.

Soon Sue and Dr. J had dinner ready so Phillip went to Patty's room and announced with a giggle that it was now dinnertime. Patty said, "OK, I'll be right there. Just let me wash up." Her hunger and eagerness to join her family sent her directly to the bathroom to wash her hands. Philip loved his younger sister, of course, but now that he became WILLING, he understood that it was an honor to have a WILLING sister like Patty.

The family had devised another Agreement to compliment the regular Three Agreements. That is, put problems away, enjoy the food and family talk. There is always a good chance a problem might get solved right there at the table without even thinking about it.

Patty took a few slow stomach breaths all the way to the top of the lungs then an equally slow exhalation. Now her mind is free of the difficult prospects of telling her classmates that her mom could not help out on Friday morning. Patty was back in the sweet spot now and again remembered that she was very hungry.

Mom served hearty vegetable soup and a tray of carrot and celery sticks, corn muffins and mom's prize winning carrot cake. When everyone was served, Philip began to give thanks. They all reached out and held hands.

His Prayer:

"Oh Great Spirit
And Creator God, we give
Thanks on this special day
For the beauty that surrounds
Us, the people here who enrich us
And the food and drink that
Nourish us. Also, please help us
Remember to serve those who are in need.
    In closing, to show honor
To unity and wholeness, we will
In silence take three slow life-giving BREATHS"

Soon everyone had completed Phillip's instruction. Thus full of ALOHA, they began to eat. Patty thought that Phillip seemed much more mature and grown-up since he became WILLING to be HAPPY. She loved Phillip's blessing. When asked, Phillip said that he read the blessing in a book about great Native Americans, and, of course, Philip added the last part about the breath.

Dr Jason is Lakota Sioux . Both his mother and his dad are teachers at the Rosebud Reservation in central South Dakota, therefore it was natural that he had many books about the culture of the native American Indians and the Lakota Sioux Indians.

Sue had been born in El Salvador. She met Dr. Jayson when, as an undergraduate, he participated in an archeological dig one summer in El Salvador. They spent a lot of time together. When Dr. Jayson went back to the University of Nebraska for the winter term they wrote letters almost every day. Dr. Jayson returned the next summer to El Salvador and there Sue and Dr. Jayson were married. The next year he graduated from the University. His degree in Archeology made him a well qualified scientist but now the family knew that Dr. J had become wiser and gentler since he became WILLING.

The delicious soup went well with the perfect corn muffins. Dr. J and Patty put butter and molasses on their muffins while Sue and Phillip mostly used butter and honey on their hot corn muffins.

Sue expressed her interest in the stack of books Phillip had lugged home. Phillip explained, "I have to write three reports for Friday so I'll be very busy doing research after school this week. Sue looked at him quizzically and wondered why he had so much work to do. Dr. J had just finished crunching his third carrot stick when he asked that very question. Phillip replied, "Dad, I have to do the reports because I agreed to do them." Dr. J did not press the issue. Besides that it was now time for Sue's famous carrot cake with the yummy frosting. Patty and Phillip looked at each other and laughed at the mention of their mom's famous cake as they rubbed their bellies and said they did not have room for more food. Sue said, "Well, we will wait a little while so the food can digest." Then Sue looked at Patty and said, "Patty, you really started something when you volunteered me to help make the museum trip possible." Patty sat at the table with a full tummy and almost stopped breathing, wondering what her mom would say next. Dr. J asked, "Excuse me, Sue. Did I miss something?" Phillip chimed in because he too was in the dark. "Yeah, Mom, fill us in," he said with a giggle. "Everyone knows Patty is so smart she will probably be the Mayor of Lincoln before she's twenty".

Patty had never thought much about being twenty and besides what's the big deal about being mayor? And what does a mayor do?

Before Patty could respond, Sue quickly filled in the blanks for Dr. J and Phillip, Sue reported the story of the dog and cat and the domino effect on the bus driver's wife and the class and Patty's possible solution. Sue said after talking to Patty the phone began to ring off the hook. Four different parents of WILLING classmates of Patty's called to volunteer for different parts of the day thus liberating the bus driver for the trip to the museum on Friday. Each parent remarked how their home life had improved since they had been taught to be WILLING. Because their child was a classmate of Patty's, they had been taught to be WILLING. So it was the WILLING parents of Patty's classmates who had earlier become WILLING that answered the call for a solution to the museum trip. Patty was relieved that she did not have to say anything more than the good news: that the trip would happen now that some parents of the class had helped solve the problem.

Phillip finished the three essays for English class and handed them in on Thursday but not before he included a short paragraph in the middle of the last page explaining to the teacher what happened with the two boys and their threat to fight Phillip and steal his books if he did not write the papers. When Friday came, the English teacher had the time to grade each paper and wanted to talk to Phillip after class. Mrs. Lui said, "Phillip.""Yes, Mrs. Lui," Phillip said remembering to breathe. "I am very much impressed with your writing abilities demonstrated in the three theme papers that you prepared. I am going to transfer you to advanced writing class starting Monday. As for Larry and Simon, I will meet with them after school today. What you did took more courage than fighting with these boys and you helped yourself by doing extra work for the experience."

Mrs. Lui sent a note around to Larry and Simon to meet her after school. Sure enough Simon and Larry showed up after school for the meeting. Mrs. Lui asked the boys to sit down and immediately began talking about how good their papers were. She went on to say that she would transfer the two boys to advanced writing class starting Monday morning. Hearing this, the two looked at each other and then back at Mrs. Lui with open mouths, not knowing what to say. Mrs. Janet Lui did not wait for an answer. She sought the TRUTH and we all know "who" that is. After a couple of questions pertaining to the papers, it was obvious the two knew nothing of the contents and then they began to speak the TRUTH. Simon said he did not think he was ready for advanced writing class and he admitted that Phillip had written the paper under threat of a fight. Larry said the same was true for him and that he never thought Phillip would write such a good paper The two admitted they were wrong to do what they did because they were just plain too lazy to do the work themselves. They both wanted to stay in general English in order to learn how to write like Phillip. They also said they would apologize to Phillip and also ask about how to be WILLING because it obviously made Phillip a happy person. Mrs. Lui told the boys that she would call their parents to inform them of what had happened. She told the boys they still owed a theme paper which by now was late. That evening after school, Phillip walked home, thinking about how much fun he had keeping the BREATH flowing and maintaining ALOHA in his thoughts. He realized that he had many more friends than ever before and furthermore, all the adults seemed to be quite friendly. This made him feel good and he always had a good time. Phillip felt grateful that he had the courage it takes to be happy. On his way home, Phillip stopped by Mrs. Gramet's home to see if she needed any chores done since she was an older person who lived alone. She sometimes needed help to lift or carry something. When Phillip came home, he carried out the trash, set the table and went to work on his homework.

By now Dr. J has realized that his family is happier, more relaxed and more creative than ever before. Limiting the TV watching and muting out all commercials (which are non-ALOHA in content anyway) has helped the children keep a wholesome, healthy, happy ALOHA mind. He thought again of that very lucky day when José and he found the cave and also the day he met the wise old Peruvian woman who threw the Bones and told him of a future he did not really believe possible.

Dr. J opened his mail and there among the letters came an invite to go to Australia for a dig in the summer This excited Dr. J because the whole family could go, but that is another story.

So in closing, let me say again, use your courage on the road paved with a full slow breath and exhalation, use the bridge (that's you) called TRUTH and keep sweet ALOHA as a constant companion and of course keep the Three Agreements as your guide. Your life will be good and the world be a better because of you. You will be happy. May the spirit of ALOHA be with you always. You can be happy when you become one of the WILLING. ay you, your family and your friends become WILLING and attain full happiness.

ALOHA from the future.

ALOHA to the past.

ALOHA in the present, BREATH by BREATH.

This is my message from the land of the WILLING. 48,004 B.C.

Chapter Twenty Eight

THE DREAM

Having finished the book, Dr. J laid it back on the table. The next thing Dr. J remembered is Sue nudging him to get up and to bed. "Remember" Sue said, "You have an early departure time for Peru in the morning.

Dr. J stands up facing the fire and while stretching his back said, "Oh, I must have dozed off after reading that wonderful book, there on the table."

"Dr. J, you must have been dreaming." Sue said with a smile, "There is no book on the table."

Dr. J turned around while rubbing his eyes to adjust to the now very dim firelight and, seeing no book, realized it was just a dream, or was it, for the next morning in the waiting room at the airport Dr. J read with interest a newspaper report from Peru about a shallow earthquake.

You will decide your future-

May it be a HAPPY one

Part Two

Epilog One

PATTY SPEAKS

Not included in the main text are some poems and comments from the family. First to speak is Patty. Remember!!! The 'happy' feeling will surprise you even after a couple of days of deep, slow, conscious breath practice. You'll giggle more often, and that's a good thing. Here's something the family discovered while learning to be 'willing'. I think this is important. Here are more guide-posts that will ensure safe passage on the road to 'happy'. I did not create this poem by myself; Mom (Sue) helped me on certain parts. Dad (Dr. J.) liked it so much he suggested the poem could be a lead-in to a series of healthy cartoons for kids like me. I think it's a powerful idea for those whose time has come. Well, here it goes: Oh, I hope you learn to sing it as I do, and again, Mom helped make up this powerful poem-song. It's called, Feel Fine.

Feel Fine

Keep the agreements

Refrain from the do-not's

Remember to breath every time

Having done this

You will do well

And much even more

you'll feel fine

Do not get angry

For anger's an ox with no cart

Do not be cynical

For cynic's are

Neither funny nor smart

Do not do gossip

For gossip

Is wrong from the start

So speak the truth only

And make the true tone be one

That comes straight from the heart

Joseph Henry Wilkinson

So let us add these do-not's to our moral compass. This helps clear the picture in daily life as how to be. Mom's got a gift for writing poems. She's the one who came up with the only-tone-be rhyme. I love my mom so much. She brought me into this world and has cared for me all of my life. And now she's teaching me poetry. Poetry may rhyme with an efficiency of words to make a point. There can be a rhyming scheme although poetry doesn't have to rhyme (but I like it to). Non rhyming poetry is free verse (that's good too) and it can be very profound and rewarding in its body of work. So let us say a little more about the do-not's. First of all, anger is an emotion that can be wiped-off the face of the earth. If anger is caught in time, "non-Aloha emotion" won't even have a chance to enter the realm of reality. Just go to the breath. Really send your mind to the stomach. Fill your stomach with air, fill your lungs with air right to the top, then at a slow pace, hold, and then slowly release the breath. After four or five of these the anger won't have the power to hang around. If one still feels it, breathe some more slow ones. Think of all the domestic violence that could have been prevented had the kids taught the parents about the conscious breath. We can do it. We know how to make the quality of life better for the entire planet and all people who live here. If you never allow anger to over take you or if you catch anger (by using your wisdom and courage to recognize what harmful effects it will have) and just breathe it away you will have a special heir of calm that others will see and admire.

Now let's address the part in the poem about the "cynic". You may not know what a cynic is. Well, the definition given in the dictionary says that a cynic is a person who believes that all people are motivated by selfishness, as you see; a cynic does not understand Aloha. The same goes for teasing someone in a hurtful manner; or just plain, hurting their feelings. When this is done, the first agreement goes out the door and the moral compass is now in the dark. Remember and respect the power of the spoken word. Our words go out to the entire universes to hear first, and before the person you are talking to even hears it. Keep your words in the 'loving' and respectful 'pink light'.

The last 'do not' mentioned in the poem is about gossip. Gossipers need practice to keep up the gossip. Gossipers sometimes practice so well that they cannot think of anything to talk about except gossip. Gossip is about somebody else and not necessarily about the truth. Gossip is a destructive force that has no fruit.

A wise man once said that there are three things that people talk about: People talk about other people (they gossip). Remember, the people who gossip do it because they practice. It takes work, and it's not much fun. People talk about things, such as washing machines, cars, and lawn mowers. People talk about ideas. (Gotcha!!!) This is evidence that the creative spirit has been dancing. People even sit down and talk about their own creative ideas, such as recipes, gardening tips etc. the list is endless. For those people who talk about things; things are just part of life. All people have to deal with things. Oh! And I might add, that it's best to have Aloha in your breath when dealing with things. When talking about ideas, or the creative aspect of ourselves, there is no limit to the wonder and grandeur of the creative Spirit. The poetry I wrote was an example of this. Oh, a great key that I found, and I think you will find is priceless as well.

This is the key to always finding your center, your sweet spot, the place from where it is easy to smile. I cannot tell you where yours is, I know where it is for me, and it's probably the same place for you. Do this and you will find the place where you can smile and feel fine. Before a meal or snack, pick up your bowl or plate of food away from the table, stand, face the East if possible, and feel gratitude, real gratitude, for the food you are about to consume. After a moment you will feel the spot in your body where you can always go and smile easily every time.

Thanks for reading my comments. There was not enough room in the book for what I had to say. See you in Australia.

Loving Light,

Patty

Epilog Two

PHILLIP SPEAKS

Here are some words from Philip, plus a poem that he just finished before printing time.

Patty writes great poetry. I'm glad she has a fine sense of humor. She will be Mayor some day. Patty never mentioned the idea of breathing loving pink, and then directing it to various targets, so here are the subjects we chose for five rounds of eight-fours. I might say here that, the reason we chose pink is because it is the color vibration closest to "love". The first breath is done by breathing pink through the heart and into the lungs. With the mind's eye, you see your self all pink, from head to toe, beautiful, loving, and harmonious in it's sweet vibration. It is your most important duty to love your self and know that as you continue, you will find more and more ways to be grateful for life, and grateful for being happy, and thus love your self even more. Remember, the more one loves 'self', the more one can love that which is, that which includes everything in our Universe, and of course, it means the neighbor's dog that barks too much on those full moon nights.

The second eight-four breath, breathe pink on the place you live, that is, your home and your community close by. The third eight-four breath, breathe pink on the Pentagon. You will see in my poem that a lot of visuals are possible. The forth eight-four breath, breath pink on Washington D.C. (Dad wrote a poem about that part). The fifth eight-four breath, breathe loving pink on the Middle East. They need our help to stop the fight. I realized that all those who are willing could do our conscious eight-four breathing at 7am and 9am every morning. Perhaps we could promote solutions that do not include unabated fear which promotes over reaction, human suffering, and human death (but I don't want to dwell on that any longer). My aim is to produce healthy solutions, to produce love and peace for our Mother Earth which is 4.5 billion years old. Most people live less than 100 years. So let us go and have some fun with my poem that I call, Re of Defense. Oh, by the way, Dad gave me the idea but I wrote most of the words. So here goes!!!

Re of Defense

We had a Civil War
And ever since,
We want-ed peace
And harmony hence.
Twice a day we pink the air,
And also remember to pink the chair
Of the one in which sits
The Secretair.
Re of Defense
We had a Civil War
And ever since,
We want-ed peace
And harmony hence.
We pink the doors
And all of the knobs
With loving, sticky
Pretty pink globs.
CHORUS
We pink the furniture
In all the rooms,
With the speed of
Multiple-sonic-pink booms.

CHORUS

We pink the mess-hall

Where there's food to eat,

And when we pink the drinks

We try to be discrete.

CHORUS

When the lights go on

In the Pentagon, we pink the air

And also remember, to pink the chair

Of the duly appointed secretair

Re of Defense.

Oh, by the way, I thought I should write a brief history of
the Civil War, (1861-1865), but before I do, I need to say
that over six hundred thousand men died in that war while
fighting for their cause. I am humbled to think of the
courage, honor and strength of these men on both sides,
blue in the North and grey in the South. I'm not a historian
but I feel I must tell you the impression I got from my
research. Seven Southern States walked out of Congress in
March/April, 1861. Because the South walked out, and
would not return, it meant the end of the United States,
which was no longer united. Under the Constitution
Congress could not meet again unless all State
Representatives were present. I must say, that the role of the
Federal Government was one of little power, as they simply
pushed paper around and made sure of good relations with
other countries (diplomacy). Every State had very strong
State Rights and every citizen had powerful Civil Rights

(I'm not talking about those referred to as slaves; they were not considered citizens at the time). It turns out that the South walked out of Congress with good cause. The South produced large amounts of cotton (with the slaves doing most of the labor) that was being sold to England; who by that time had a billowing steam engine industry, the English could pay big bucks for a bail of cotton, but a few Northern big business men (called hereafter, Mulas') wanted to buy cotton from the South at a much reduced rate. The Southern plantation owners could not see any advantage to selling at home for a smaller price. So the 'Mulas' talked to the paper pushers (read Federal Governments) into placing a high tax (tariff) on all cotton shipped abroad. The South knew that this was unconstitutional on the face. The Feds were not to get involved in the affairs of Commerce (business), but they did, allowing the Mulas' to influence unconstitutional rules (mainly in the Mula's favor and self interests). This became the first step toward incredible human suffering for all of those people living south of the Mason Dixon line and east of the Mississippi River and also the Northern soldiers, many of whom were drafted, thus being forced to go against their will and to fight and maybe die.

So there you have it! The Civil War was fought because the Mulas' said that the price of cotton was too high. That's it. Hundreds of thousands of men, young and older, died on both sides. These men were robbed of their right to have a full life and a family. Robbed of memories yet to be made because their life was cut short due to the price of cotton. It should be noted, that if a soldier was injured in battle he would probably die because medicine was not advanced enough to address gun shot wounds, burns and ensuing infections.

Keep in mind, the war was not about slavery. Both the North and South had slaves; Slavery was tossed in later as a reason for the war. It was like the pot calling the kettle black (no pun intended). Lincoln, by making slavery an issue, gave the Northern troops something else to fight about, since the price of cotton was a poor reason to go die somewhere down South. But I'm getting ahead of myself. When the South walked out no one knew what to do, no one except the Mulas' who saw a chance to create chaos. Only two weeks passed and the North was up and running again. Lincoln was declared President of the union, and after placing troops with guns pointing out from D.C. for two weeks, they declared Martial Law (suspension of all Constitutional rights) which by International Law, meant the North could now attack the South and try to convince them to reunite with the North. To my knowledge, and to this day, we are still under Martial Law. The Mulas' got their way 'big-time'. Too bad the Mulas' didn't know about Aloha, or the agreements, or the do-not's, or even how to breath properly. If we could go back in time, say, 1858, (the War started in 1861), we could teach every one around us to 'pink' D.C. What might have resulted could go like this: Lincoln could have said, "Oh Well, Let's just scratch those Southern States off the Constitution, and let's get on with governing what we have left". Simple. The South would have their group of States the Southwest would have theirs. Texas could stand on its own. The Northwest could have its group of five or six States, thereby smaller and more manageable countries would have smaller Federal Governments, and State rights would be strong again. There would be no super military Power. After all, super powers are driven by fear.

Here's a piece of information that I read: In August, in the year 2001, the then Secretary of Defense admitted that, due to poor bookkeeping, the military lost two trillion dollars. When stuff like this happens, I say 'down size'. Abandon the Pentagon and turn it into a homeless shelter. I know that sounds radical, and possibly if I read more I would change my opinion. But I do know this; I know that when power, be it military, or money held by a few is moved without love, it is unfocused in how to use it's power properly. Note: nowhere in my research did I find where the common man had any say about the War. At least 200,000 Northern men did not feel the need to kill Southern men, thus they resisted the draft, but regardless they were forced to go to war. Certainly, common sense and natures way were not served. OK ! If you're wondering where I'm getting my information, I can tell you that since I only watch one hour of T.V. a night, usually PBS news, I have plenty of time to read. Recently, I've been reading about the way a super military power can create a Civil War. There is much to learn from the past. Historians are great writers and they do boundless research. It is a joy to read well researched and written material. I am so happy, so happy that I know the basic guide post that led to the place that we have come to call happy. Perhaps after you read my Civil War research you can appreciate my poem even more (thank you for allowing me to be long winded on this issue, but I thought it needed to be said). About this poem; I hope you liked it and had some fun. I hope you take the poem to heart. We can do our part to change the world. When you hear the word 'love', think pink, in order to get closer to the vibration of wholeness and unity. In other words, we are all in this together. It is time to teach each other how to think pink.

Don't ask me where the words come from that are found in this poem. When you calm your mind long enough, you too can dance with the Creative Spirit. My poem would not have happened had I been watching T.V. with all the junk-like commercials. I allowed the Creative Spirit, the one which all of us have access to, to dance with me while I compose my poem (although Mom helped me a little bit). See you next time in Australia.

With love and pink light,

Phil

Epilog Three

DR. J. SPEAKS

Dr. J.

I am thinking about creating cartoons in which the willing concept is portrayed in an enlightening way. We may very well help people who have had anger attacks before understanding the responsible breath. Breath first, asks questions later. Think about the fights that didn't happen because someone remembered to breath. We must grow as people willing to breath right through psychic intrusions of a negative nature. We can preclude arguments that could have possibly started over NOTHING. We can alter certain behavioral patterns. It's OK to remind a friend to breath, because we are all keeping Aloha, day by day, breath by breath. It is the way to be in support of our rights, to be happy as often as possible. However, if we break an agreement now and then, we should recognize it as soon as possible and clean it up and forgive ourselves and then breath on. We will find that it is a good idea to hang with others who are willing. They can catch slip-ups and support you on your way back to the sweet spot, the true center. It is very important to hang out with positive, happy people. If you are with negative people, they may rub their negativity off on you.

I have written a poem to illustrate the power we have as a people using the breath, visualization and intention to help bring our Nation back to the joyful balance of peace and harmony using pink as our conduit to success.

My poem is directed toward the idea that we can actually 'clean up' Washington D.C. This project, of course, would take serious focus and daily repetition to complete the task; as outlined by Philip, using the five eight-fours at 7am and 9am. My poem is called: Washing D.C. So here goes!

WASHING D.C.

I'm breathing now, don't bother me,

I'm in the process of washing D.C

They say D.C. has a little dirt,

So a good clean washing surely won't hurt.

I'm breathing now, don't bother me,

I'm in the process of washing D.C.

I send some rain, pure enough to drink,

And watch it turn the House from white to pink.

I'm breathing now, don't bother me,

I'm in the process of washing D.C.

They say law makers could use a lift,

So I'll send a pink snow storm, if you get my drift.

I'm breathing now, don't bother me,

I'm in the process of washing D.C

I see the High Court, Dressed in matching blacks

So I'll pink their robes Since they don't wear hats

I'm breathing now, don't bother me,

I'm in the process of washing D.C.

I see the vibes changing to pure love

If it's pink below, it's for sure above.

I'm breathing now, don't bother me,

I'm in the process of washing D.C.

I see the children playing, laughing with glee,

So I'll gladly continue washing D.C.

I'm breathing now, don't bother me!

When we teach others how much fun it is to be willing, they will join into the fun. When we hang out with negative people, who have negative thoughts, and a negative understanding of the moral compass, then, believe me, it will rub off on us. Hang out with company who are willing, there will then be a small chance of any slip-ups. Follow this moral compass laid out in Patty's poem and teach others. Demonstrate to others your responsibilities. Match your words and don't give in. When others see that you are always on time, they will respect you because by your being on time, shows respect for others. To be prompt you must arrive early, this will insure that you are matching your responsibility. There was not room in the first part of the book to speak to this, almost scary, concept called 'responsibility'. Now I want to say something about my Lakota Sioux ancestor's history of the near past (2-3 hundred years).

All Native Americans have one thing in common. What do you think it is? Your right! Of course! The answer is Mother Earth.

All the inhabitants in this land before Columbus were well aware that all of life comes from Mother Earth. The Native American understanding of that which is, is illustrated in ceremonies where the great Spirit and Mother Earth are acknowledged and respected with gratitude.

What would life be like without the Great Spirit and Mother Earth? Well, there would be no gifts from above. There would be nothing to stand on. Simply put, but just the same, imagine yourself as an observer somewhere in the Dakotas in the year 1100 A.D. The people of that time who lived there had no chance of hearing about different religions that could attempt to explain a view of God. These people knew what was real for them. They knew that Mother Earth provided rain, good hunting and good weather. For that they were grateful. They celebrated with joy and laughter in times of abundance. To say it another way, "Those hats which love to travel are grateful for the heads they sit upon".

If a Lakota Sioux followed his/hers ancestors back to fifteen grandfathers, then one would discover that they've always lived in that area of the world. How do you suppose they felt about Mother Earth? Without Mother Earth there is no life. I can feel my ancestors close by as I write this. This was the truth for the Lakota Sioux who lived before those gold-diggers who arrived in the late 1800's. It was the truth before gold was discovered in the Black Hills of North Dakota. The Black Hills were considered sacred to the Lakota Sioux'.

After many battles and brutal hardship the proud Lakota Sioux' were placed on reservations. They were forced to live in places not of their choosing. If this is not 'unconstitution-al' then I don't know what is.

How could the United States be party to such a travesty of injustice? It sounds like Martial Law was still in force going back to the Mulas' and the Civil War (well, I got a little long-winded on that part of the truth).

Let's say that you, along with a plastic bottle of drinking water need a ride to town and a friend picks you up. As you ride in the back seat of the car you suddenly realize that you are getting too hot, so, you roll down the window to let in air. Now, when you arrive at your destination, do you jump out of the car and go about your business? Or, do you roll up the window and pick up your empty plastic drinking bottle off the floor and say thank you?

Now, here's a riddle for you. What is the very first thing that you should do when you enter this room and you do it last thing when you exit a different room? In every kitchen of every household, there should be a rule (if there isn't one already) to: "Leave it as you found it, or leave it better than you found it". That's common sense and it's responsible. When you always 'leave it as you found it', you will gain respect from others, and what's more, you will respect your-self even more. So, take care of the little responsibilities and the big ones will be easy. As to prompt you a little more, let's go through the 'wake-up' call to get you out of bed in the morning. Alright! You go to the bathroom and do your thing. Now, what are the details? Well, when you wash yourself, feel the gratitude for your body that carries you around. Feel the gratitude for your teeth as you brush them; and remember this: When you consciously acknowledge the teeth as the servant for life, then you are strengthening the life-force in your teeth; after all, you might live to be 145, or more.

The answer to the riddle is not turn on or off the light, but to wash your hands when you enter the kitchen, and when you exit the bathroom. Back to your room; your dirty cloths are in the middle of the floor, 'pink them up.' Clean your space. Make it tidy. After all, it's where you live. Show yourself love by cleaning up your space. When your eyes fall on a messy room then it is a 'non-Aloha' room. Use your creativity to rearrange your room and move out stuff from view that you haven't used in the past six months. Do other things that may not be your responsibility and make it your responsibility, such as: washing a stray dish, caring out the trash, sweeping the floor, you will then feel the joy of the second agreement. When you become a responsible person, you become a very powerful person. Others will recognize you for this. When people realize that you can match your word, they will want you to take leadership positions, this way you can have even more fun sharing Aloha. You will feel confident, with a moral compass to guide you. Remember, Aloha is a powerful gift. So use it, use it every day, in every way. You will realize (before you're half way through) that your life is becoming better and better. Smile a lot, breath every moment, with every smile. Give Aloha, and keep giving Aloha, so that your light can shine at its very brightest. Do not be concerned when Aloha is not returned. Remember, your only concern is to keep Aloha. Keep Aloha breath by breath, day by day.

Ones option might be to 'freeze' breath. This will only allow 'non-Aloha life forms' to enter, and you will find yourself 'down the drain' and 'down in the dumps' because you are thinking non-Aloha negative thoughts about someone who didn't return Aloha. That's an option we wisely refuse every time. We as people are potentially divine; therefore, we must be present at all times. Keep breathing. Keep the ball up in the air by maintaining your Aloha. Besides, it's so much more fun being happy. Think of people with high regard. We are all made from the same perfect stuff. We are all seeking the truth. It is only with the understanding of wholeness and unity that we are able to become the truth. Having said all of this, here is another poem that the author dedicated to sphinx, my loving orange cat whose tail I accidentally stepped on.

Kitty Becomes Cat

By: Joseph Henry Wilkinson

When does a Kitty become a Cat?

I will do my best to answer that.

But my solution may not be so pretty,

Because I am uncertain when a Kitten becomes a Kitty.

So when does a Kitty become a Cat?

Some say it happens slowly,

Others say, a little faster than that.

Oh, the life of a Kitty is a sweet one to map

They play till they're hungry then eat'n take a nap.

A life style I surely would love to adapt.

So when does a Kitty become a Cat?

Well I have wondered and pondered and thought
about that,

And now I can state the answer as a true natural fact

If you would only allow me to quickly recap.

Last night my Kitty jumped up in my lap

I looked away briefly and when I looked back,

Where once sat my Kitty , now sits a Cat.

I so loved that Kitty but so much for that.

Now I need to figure out how to make friends with
that Cat.

It's true a Cat sleeps sixteen hours a day

Thus, a twelve year old Cat has slept eight years away.

Oh, the life of a Cat is a sweet one to map

They sleep till they're hungry then eat'n take a nap,

Or look around for an uncluttered lap

Where they are sure to get rubs and pets and stuff

And they need only purr till they're sleepy enough

And when they wake up they excuse from the lap,

And do the 'cat-walk' to the place where they snack,

And since they're not sleeping they decide to take a nap,

And if you are lucky it could be you're lap.

So when does a Kitty become a Cat?

I say, "In a blink of an eye—or—maybe—a litter faster than that".

That's when a Kitty becomes a Cat.

Pure Love and Pink Light,

Dr. J.

Epilog Four

SUE SPEAKS

And now here are some words from Sue and also one of here famous poems. Remember that she wrote Ergo Life. Go back and check out the rhythm and rhyme scheme in Part I.

Dr. J. says a lot of very smart things. I'm glad he and I are friends. He writes pretty good poetry too. It's fun to write, not just poetry (although it's my favorite) but also short stories. Write letters, its fun. Here's a good idea. Write to a friend or to someone you see every day, or to a Grandparent or an Aunt, and tell them something that you like about them. Then tell them about a book you recommend and why, or tell them if you are involved in some creative project. It is good practice. To be able to express yourself on paper is a skill that can be acquired with practice. It is one of the skills a leader must have. So I suggest you practice writing. Oh, by the way, (no one has mentioned it recently) how is your stomach doing? Is it relaxed? Is it moving in and out with the breath? If not, relax your chest and shoulders, it'll kick in, but you have to think about it first. I would like to go a little further on what the family has started regarding the concept of thinking pink. Imagine millions of people thinking pink at 7am and 9am every morning. After a period of time we can change many things. We can immensely increase self love, love for the place in which we live and, any place we choose to send love will be enhanced and harmonized. Love is harder to send than pink. At first, love may be a difficult concept to send, while visualizing pink is easy. So, feel love and think pink.

While I was writing this Patty came by and said, in her idealistic way, "The lights in the Pentagon may turn pink and even after they change the bulbs the lights will be beautiful loving 'light' pink. The pink lights will make it so uncomfortable for those sitting around planning another war that they will simply walk out and hopefully find a "real job." What Patty said sparked a poem that I hope you'll enjoy. Since I have learned to calm my mind, the creative Spirit seems to be close at hand. So, here goes. I hope you like it. It has a good ending. It is called Abandoning the Pentagon.

Abandoning The Pentagon

They are abandoning the Pentagon

One by one,

They say making war

Is no more fun.

They are abandoning the Pentagon

Two by two,

Saying war is harmful

For me and you.

They are abandoning the Pentagon

Three by three,

Saying war doesn't work,

Just read your history.

They are abandoning the Pentagon

Four by four,

Saying we ain't gonna make

War no more.

They are abandoning the Pentagon

Five by five,

We gotta stop making war

While we're still alive.

They are abandoning the Pentagon by

Sixes and sevens,

With lines as long as

Ten or eleven.

They are abandoning the Pentagon

Eight by eight,

Saying since we stopped making war,

The world is truly great.

They have abandoned the Pentagon

All except one, he sits in the pink chair.

Having so much fun,

Because he is now the leader of none.

Since they abandoned the Pentagon

It's a nice to be,

For all of the homeless

Living in D.C.

I shall now present to you a revolutionary idea. If you decide that my idea is in line with the principles of Aloha, then YOU might become an Ambassador of the 'Aloha Movement'. Write poetry, or make a video to illustrate this idea, you could even write a play and present it to the local Church. It is of my opinion that we all need this information in order to grow and evolve as a society. This information is necessary to know.

Remember that we are responsible for how we feel. If something happens that hurt your feelings, just do what is right for that situation, and breath forward. If someone, or something, makes you angry or sad, just challenge your self and go back to the sweet spot. Try hard and keep doing it, think of something else. Try hard and keep doing it. Why? Because if you dwell on what has upset you, it only gives it power. So do not give power to a person, place or thing that does not support happiness. You might have great trouble trying to not think of something, so instead, think on purpose; consciously think happy thoughts. This is how one might control thought patterns but by all means refrain from non- Aloha thought patterns. Our private thoughts are precious and this we must respect, we must respect the power of thought by maintaining Aloha at all times. We are always encouraged to maintain our way to happiness. If there is bad news, or stray gossip, just act the same and stay clear of such traps. This is not what you are about. Just excuse your self from company that starts gossiping by saying, "Thank you, I must go now", and then leave. Keep informed but do not buy into it. Always try to stay calm. Reinforce all things in the light. Socialize with people who share the same values and happiness that you do. You will have so much fun, after all, people full of Aloha are just naturally full of fun, and fun to be around. Remember, all your friends and family will learn to be willing, but this will take time. Support every one who is willing, and especially support those who are 'trying' to become willing. If they slip-up, just smile; say "Aloha" in a calm, gentle voice and then remind them to breath. They will quickly and willingly comply. We are all here to help, support and respect one another.

Masaru Emoto wrote a book called, The Hidden Messages In Water. Go buy it if you can, or ask for it at a library. Emoto's book contains scientifically verified information about water. Here's what was found: No matter what environment the water is in, it will always reflect it. If water sits, even for a short amount of time (say in a jug), and is in the presence of classical music, the water will reflect the vibrations of the music. When water is frozen at -5 degrees Celsius the water will form crystals that reflect that energy. One can see this under a microscope. If water sits in the presence of acid, rock, or rap music, the crystals in the water become completely distorted. Micro waved water will show otherworldly forms. You might re-think how you heat food. Knowledge is power!

Dr. Masaru Emoto is one of my heroes in this century. He is very much full of light. He wrote a brilliant book called, The Hidden Messages In Water, and I quote:

"We're brought up believing that sticks and stones may break our bones, but words will never hurt us. Now, however, a top Japanese scientist has revealed startling findings that that show the very real power of words to hurt, heal and influence energy at the most literal level. It's a discovery that could change the way we see our world.

Using high speed photography, Dr. Masaru Emoto—whose work got wide spread attention in the film What The Bleep Do We Know?--found that crystals formed in frozen water change in the presence of specific thoughts, either positive or negative. The Hidden Messages In Water, 32 pages of full-color photos show the remarkable results: water exposed to words such as "love and Gratitude" (or other affirming energies, like classical music) shows brilliant, complex snowflake patterns, while water exposed to negative words, harsh music, or pollution forms asymmetrical patterns with dull colors. And because our body and earth consist largely of water, the implications for us are unmistakable. With passion and precision, Dr. Emoto explains his process, explores his findings, and shares his conclusion—a call to personal health, environment renewal, and peace." So in conclusion, try this: Place the words, 'love and gratitude' on the side of a container that holds your drinking water because when you drink the water, you get all the benefits of it; you will soon feel the change in yourself, and in the rest of your family. This is what I did. I bought a big pitcher and wrote, in permanent marker, 'love and gratitude' on the side of it. I drink the water, and I also use it for cooking and baking (Oh yea! If you don't have a jug, then you must be 'willing' to go and get one, soon).

Phillip said that the way to use this concept, and direct it to the highest good, is to use it in the last weeks of November and December. That's true since we have Thanksgiving during the last week of November, and so we could use this time to show, and spread, "love" and "gratitude". At the beginning of the Thanksgiving-gratitude meal, maybe we could all hold a glass of love and gratitude water up and consciously say and feel gratitude for all of life. During the last ten or so days of December we could call The Season of love and Gratitude. We might start by celebrating the Winter Solstice (the shortest day of the year) with love and gratitude, for all that has come to pass, and for all that will pass during this time of year.

Dr. J. just said, here is a good that we could add to the season of "love' and 'gratitude', as a gift giving idea, and this is how I see it: Starting with a calendar for the New Year, I see parents mapping out the next year, planning things to do, and places to go in the course of the following year.

For example:

February 24th, we go to the zoo.

April 1st, we go to the museum.

July 3rd and 4th, we go camping (we have a birthday at that time so we plan to make ice cream), in other words, make plans for gifts yet to be given and received.

Things don't always happen the way you plan it to happen but if you make a plan, then at least something could happen that's a memory maker. It is a good way to share 'love' and 'gratitude'. So get a new calendar and start making plans. It is most important to fulfill the agreements made. As I said, and despite all else, your plans may not work out. Don't worry. It's not a problem. Make a new plan. Perhaps the last day of December (Dec.31st) would be a good day to exchange calendars and give thanks to the source of all life. You could celebrate that last day of the year with heartfelt 'love' and 'gratitude' by exchanging creative hand made gifts. Well, when Phillip stopped talking (he had already said a mouthful) Patty said, "What about that commercial Santa Clause all dressed In Red?" Phillip smiled and said, "Patty, that's something we need to change. What happens between a child and his/her parent at Christmas time isn't fair".

The commercial Santa Clause has mind-controlled many parents to play the role in a game with no-winners. The parent (say of a five year old) labels the present with the words: 'From Santa!' The child opens the present, likes it, and feels gratitude toward 'Santa' (the false and invisible 3rd party in the room). Since the child did not feel gratitude toward the parent, the bond of love and gratitude was not complete when the parent handed the child the present. Because of this first lie, the child can lose respect for his/her parent in later years. Without gratitude, respect cannot increase in the early years.

If we have 'love' and 'gratitude', we have tapped into the most powerful force on Earth. A parent/child bond should be reinforced at the end of each calendar year. There should be no mixed messages here. With love, the parent gives the child a gift, making sure that the child clearly knows who the gift is from, and to whom it is addressed. A child will feel gratitude toward his/her parent and through gratitude, respect is built. By giving this "Santa" credit for something that "it" didn't do only creates domestic sabotage. Once the lie is told, license is acquired by the child to judge the parent. Also, once a lie is told it is easy to tell another lie, therefore, do not lie. Can you see how this commercially red-suited man sets 'it's' trap for all of our parents? It tricks them into telling children lies. Eventually, the children will find out the truth about this, 'red-suited Santa'. By the time children reach their teens, they have already lost many opportunities to show gratitude along with the enormous amount of respect they should feel for their parent. This 'Santa Red-Suite' should best be forgotten.

The living trees that are cut down for the Christmas season could be avoided by using long lasting plastic trees (that is, if you really think you need a tree) or a not yet invented LED lighted crystal form of 'love' and 'gratitude' could take it's place. We can do better on the end of the year celebration. We need to be more conscious of what, and why, we are celebrating.

The season of 'love' and 'gratitude', taking place between December 21st and December 31st, would be better served than on December 25th, which is too commercialized and counter-productive for those of us who are "willing". These are thoughts only of November and December; so what about the rest of the months, February, March and October? Valentine's Day is a perfect Pink Day to celebrate 'love' and 'gratitude'. It's on this day that we can express our feelings of 'love' and 'gratitude' not only toward our teachers, class mates, parents, friends and family, but especially to the creative Spirit that is alive in all of us.

In celebration we might write creatively, or even make something creative, but what ever we do we have to put forth effort to create a space for the magical energy to speak through us. Halloween is strictly commercial. We could take this time to celebrate the harvest of food in abundance, and to show 'love' and 'gratitude' toward those who have passed before us. Spend solemn time, and festive time looking at photographs of loved ones who have passed through the portal of death. It's good to remember others if you wish to be remembered by others.

It is important to have a well exercised and muscle toned body. Patty and Phillip have incorporated daily exercise as a means to get the heart pumping for at least 20 minutes. Jump rope is a good one, jogging or aerobics. You could play soccer or basketball. Whatever works. Why? Because it's healthy. It tones up the muscles, it keeps you in tune with your body, and most importantly, it will allow you to be more connected to your breath. Also your grades will improve in school, and that's a good thing.

So you see? Our quest for happiness is never ending. We must eat healthy food, fruit and vegetables; you may be surprised that when you get used to eating good, healthy food, you will lose your taste for foods that over stimulate the taste buds on your tongue. Healthy snacks too, such as dried fruits and nuts are good for overall health. Drink plenty of water that has been charged with 'love' and 'gratitude'. Refrain from eating sugary, salty foods, or drinks, that may be full of preservatives, this is not health food for us. Remember the statement by the 'willing' master. "You are potentially divine and with your understanding of wholeness and unity become the truth".

Oh, I almost forgot. A communal, socialist type of

government could evolve as the perfect model for citizens in their quest for happiness. The model may prove to the rest of the world to be superior to capitalism in every way. Rich people will not like it until they find out the truth. Happiness becomes the ultimate goal for all people living on Mother Earth. Thus came the end of war, violence, anger and hatred. Also, I almost forgot this important piece of advice pertaining to the spoken word. When I say anything, no matter what it is, the power of my words change the universe, and the same is true for you. Therefore, it is a good idea to clean up our words. Refrain from "non Aloha" words, terms or concepts. Don't even go there and they will drop away , and will no longer be part of you. You may slip a time or two at first, but just forgive yourself and an go on. It takes courage to make changes. Your speaking will have grace when you refrain from introducing negative patterns with your spoken word. Therefore we should learn to speak in pure loving tones about our goals. That is to say, do not give energy to an unjust reality, such as might be imposed by a government that is out of balance. I would like to share something with you that I learned from Neelie, a psychic friend. Although there has been much cause for alarm throughout the planet on nearly every subject matter pertaining to the human race, these events and dramas that are playing out are merely setting the stage for a far greater event to come: The merger of your world with ours. If this does not seem possible to you, given the amount of darkness that still appears to exist on this planet, consider this: God the Creator, is the source of ALL there is. The darkness is not a result of the absence of the source; for in truth, within all darkness there is source. I say this because without source nothing would exist, not even the darkness. Source created both darkness and light. Source separated the light from the darkness but the darkness was still His creation, was it not? The key is that source created the law to rule both darkness and the light, and that law is: FREE-WILL CHOICE. All of creation is subject to this law, even those in the Angelic realm. The choice is fundamentally this: Do you choose to create with the energy of unconditional love...(ALOHA) or do you choose to create from the negative emotions, of fear, guilt, and the need to dominate and control? In order to better to understand the result from creating with negative emotions, source, allowed the darkness to exist and grow in power throughout the Universe. From the beginning of creation all beings were given the power to choose, the Archangel Lucifer who was the brother of the Angels chose to play the part of darkness and become a master of the dark energy. He recruited many other angelic beings to join the dark forces, and they have don a stellar job in gaining new recruits, have they not? But it is only a matter of time before the forces of light will be able to "divide and conquer" the dark forces. This will usher in an era of light upon this planet that will allow our worlds to merge.

The family is looking forward to our next book and the surprises that are in store for us in Australia. Good Ai' Mate,

With Love and Gratitude,

Sue

About the Author

Joseph Henry Wilkinson was born in Guymon, Oklahoma, on March 31, 1940. He is primarily a science teacher, inventor, and poet. Mr. Wilkinson, a social activist, moved to the Hawaiian Islands in the summer of 1985 and became involved in the Hawaiian Sovereignty Movement. He has made his home on the big Island of Hawaii in the shade of the largest active volcano in the world. He continues to be involved in social activities within his community and it is through his love for teaching and his dedication toward the children of our future that he writes this book.

For more information please visit us at:

www.the-willing-spirit.org

See 1stWorld Books at:

www.1stWorldPublishing.com

See our classic collection at:

www.1stWorldLibrary.org